Angelika Keller

Synthesis and enzymatic testing of reversible terminators for SBS

Angelika Keller

Synthesis and enzymatic testing of reversible terminators for SBS

Südwestdeutscher Verlag für Hochschulschriften

Impressum/Imprint (nur für Deutschland/ only for Germany)
Bibliografische Information der Deutschen Nationalbibliothek: Die Deutsche Nationalbibliothek verzeichnet diese Publikation in der Deutschen Nationalbibliografie; detaillierte bibliografische Daten sind im Internet über http://dnb.d-nb.de abrufbar.

Alle in diesem Buch genannten Marken und Produktnamen unterliegen warenzeichen-, marken- oder patentrechtlichem Schutz bzw. sind Warenzeichen oder eingetragene Warenzeichen der jeweiligen Inhaber. Die Wiedergabe von Marken, Produktnamen, Gebrauchsnamen, Handelsnamen, Warenbezeichnungen u.s.w. in diesem Werk berechtigt auch ohne besondere Kennzeichnung nicht zu der Annahme, dass solche Namen im Sinne der Warenzeichen- und Markenschutzgesetzgebung als frei zu betrachten wären und daher von jedermann benutzt werden dürften.

Verlag: Südwestdeutscher Verlag für Hochschulschriften Aktiengesellschaft & Co. KG
Dudweiler Landstr. 99, 66123 Saarbrücken, Deutschland
Telefon +49 681 37 20 271-1, Telefax +49 681 37 20 271-0
Email: info@svh-verlag.de
Zugl.: Frankfurt/Main, Uni, Diss., 2009

Herstellung in Deutschland:
Schaltungsdienst Lange o.H.G., Berlin
Books on Demand GmbH, Norderstedt
Reha GmbH, Saarbrücken
Amazon Distribution GmbH, Leipzig
ISBN: 978-3-8381-1355-5

Imprint (only for USA, GB)
Bibliographic information published by the Deutsche Nationalbibliothek: The Deutsche Nationalbibliothek lists this publication in the Deutsche Nationalbibliografie; detailed bibliographic data are available in the Internet at http://dnb.d-nb.de.

Any brand names and product names mentioned in this book are subject to trademark, brand or patent protection and are trademarks or registered trademarks of their respective holders. The use of brand names, product names, common names, trade names, product descriptions etc. even without a particular marking in this works is in no way to be construed to mean that such names may be regarded as unrestricted in respect of trademark and brand protection legislation and could thus be used by anyone.

Publisher: Südwestdeutscher Verlag für Hochschulschriften Aktiengesellschaft & Co. KG
Dudweiler Landstr. 99, 66123 Saarbrücken, Germany
Phone +49 681 37 20 271-1, Fax +49 681 37 20 271-0
Email: info@svh-verlag.de

Printed in the U.S.A.
Printed in the U.K. by (see last page)
ISBN: 978-3-8381-1355-5

Copyright © 2010 by the author and Südwestdeutscher Verlag für Hochschulschriften Aktiengesellschaft & Co. KG and licensors
All rights reserved. Saarbrücken 2010

Meiner Familie gewidmet

"Of the three main activities involved in scientific research - thinking, talking, and doing - I much prefer the last and I am probably best at it. I am all right at the thinking, but not much good at the talking."

Frederick Sanger (*1918), biochemist

Table of contents

1 The invention of new sequencing technologies 7

1.1 From first- to second-generation sequencing............................7
    1.1.1 Historical background...7
    1.1.2 The need for second-generation sequencing.......................9
1.2 Sequencing by hybridization 10
1.3 Pyrosequencing.. 12
1.4 Sequencing by synthesis 16
    1.4.1 The invention of reversible terminators for SBS 16
    1.4.2 The array-based SBS technology 20
    1.4.3 The EU-project "ArraySBS" and its aim 24

2 Development of an array-based SBS method 27

2.1 Selection of an appropriate polymerase for SBS 27
    2.1.1 Function and properties of a polymerase............................ 27
    2.1.2 Polymerase selection with unlabeled reversible terminators................. 32
2.2 Polymerase acceptance tests 34
    2.2.1 Materials and methods... 34
    2.2.2 Incorporation of 3'-O-CEM-dTTP.. 35
    2.2.3 Incorporation of 3'-O-CE-dTTP... 36
    2.2.4 Concluding remarks 37

3 Goal of this PhD thesis 38

4 Synthesis of reversible terminators for SBS 40

4.1 Retrosynthesis of the four 3'-O-modified key compounds 40
    4.1.1 Retrosynthesis of the key compound for the A-terminator 40
    4.1.2 Retrosynthesis of the key compound for the G-terminator 41
    4.1.3 Retrosynthesis of the key compound for the C-terminator 43
    4.1.4 Retrosynthesis of the key compound for the T-terminator 44
4.2 Strategy for selective 3'-alkylation of 2'-deoxyguanosine....... 45
    4.2.1 Dialkylation of partially protected 2'-deoxyguanosine...... 45
    4.2.2 Selective 3'-alkylation of fully protected 2'-deoxyguanosine 46
4.3 Synthesis of the 3'-modified key compounds 51
    4.3.1 Synthesis of the pyrrolo[2,3-d]pyrimidine moiety: Strategy 1 51
    4.3.2 Synthesis of the pyrrolo[2,3-d]pyrimidine moiety: Strategy 2 54
    4.3.3 Synthesis of 4-amino-7-[2-deoxy-β-D-*erythro*-pentofuranosyl]-5-iodo-7H-pyrrolo[2,3-d]pyrimidine 55
    4.3.4 Synthesis of 4-amino-5-[3-amino-prop-1-ynyl]-7-[3-O-(2-cyanoethyl)-2-deoxy-β-D-*erythro*-pentofuranosyl]-7H-pyrrolo[2,3-d]pyrimidine 58

4.3.5	Synthesis of 2-amino-7-[2-deoxy-β-D-*erythro*-pentofuranosyl]-5-iodo-*7H*-pyrrolo[2,3-*d*]pyrimidin-4-one: Strategy 1	61
4.3.6	Synthesis of 2-amino-7-[2-deoxy-β-D-*erythro*-pentofuranosyl]-5-iodo-*7H*-pyrrolo[2,3-*d*]pyrimidin-4-one: Strategy 2	65
4.3.7	Synthesis of 2-amino-5-iodo-7-[3-*O*-(2-cyanoethyl)-2-deoxy-β-D-*erythro*-pentofuranosyl]-*7H*-pyrrolo[2,3-*d*]pyrimidin-4-one	69
4.3.8	Synthesis of 5-[3-amino-prop-1-ynyl]-3'-*O*-(2-cyanoethyl)-2'-deoxycytidine	71
4.3.9	Synthesis of 5-[3-amino-prop-1-ynyl]-3'-*O*-(2-cyanoethyl)-2'-deoxyuridine	75

5 Monophosphates as model compounds 77

5.1	Cleavage experiment in a heterogeneous system	77
5.2	Synthesis of 3'-modified monophosphates	77
5.2.1	Synthesis of 3'-*O*-(2-cyanoethoxy)methyl-2'-deoxythymidine-5'-phosphate	77
5.2.2	Synthesis of 3'-*O*-(2-cyanoethyl)-2'-deoxythymidine-5'-phosphate	83
5.2.3	Synthesis of 3'-*O*-(2-cyanoethyl)-2'-deoxyadenosine-5'-phosphate	88
5.3	Cleavability of the 3'-modified monophosphates	91
5.3.1	Cleavage of the CEM function using 3'-*O*-CEM-dTMP	91
5.3.2	Cleavage of the CE function using 3'-*O*-CE-dTMP	94
5.3.3	Cleavage of the cyanoethyl function using 3'-*O*-CE-dAMP	97
5.3.4	Cyanoethyl cleavage on an oligomer	100

6 Summary 103

6 Zusammenfassung 108

7 Experimental part 113

7.1	Chromatography	113
7.1.1	Preparative Column Chromatography	113
7.1.2	Thin Layer Chromatography (TLC)	113
7.1.3	Fast Protein Liquid Chromatography (FPLC)	113
7.1.4	Buffers and methods for RP-FPLC	113
7.1.5	Reversed-phase High Performance Liquid Chromatography (RP-HPLC)	114
7.1.6	Buffers and methods for RP-HPLC	115
7.1.7	Anion-exchange High Performance Liquid Chromatography	117
7.1.8	Buffers and method for anion-exchange HPLC	117
7.1.9	Nuclear Magnetic Resonance (NMR) Spectroscopy	118
7.1.10	Mass spectrometry	118
7.1.11	Elementary analysis	118
7.1.12	List of chemical reagents	119
7.1.13	List of synthesized compounds	124
7.2	Synthesis and analytical data of all compounds	128
7.3	Oligonucleotide Synthesis	212

8 Annex 213

8.1	NMR and mass spectra	213
8.2	Abbreviations	218
8.3	Literature	221
9	**Publications and Presentations**	**230**
9.1	Publications	230
9.2	Posters and presentations	230
9.2.1	Posters	230
9.2.2	Oral presentations	230

1 The invention of new sequencing technologies

1.1 From first- to second-generation sequencing

1.1.1 Historical background

The DNA structure was discovered in the early 1950's by Watson and Crick[1a] who described it as follows: The deoxyribonucleic acid is a double-stranded polymeric molecule with a phosphate-deoxyribose backbone consisting of the four nucleobases 2'-deoxyadenosine, 2'-deoxycytidine, 2'-deoxyguanosine and 2'-deoxythymidine. Today it is known that DNA stores the genetic information of every living organism and is therefore the blueprint of the phenotype of any individual. Only certain sequence regions of the DNA are transcribed by enzymes into messenger ribonucleic acid (mRNA), which is a very similar, often single-stranded polymeric ribonucleotide (see figure 1). The mRNA serves as matrix for the translation into the corresponding amino acid sequence that forms a particular protein depending on the sequence of the mRNA. Not every sequence of the entire genetic code is transcribed and translated, which means that the expression of the genetic heritage is regulated by epigenetic processes like e.g. methylation of DNA, paramutations[1b] and gene silencing[1c].

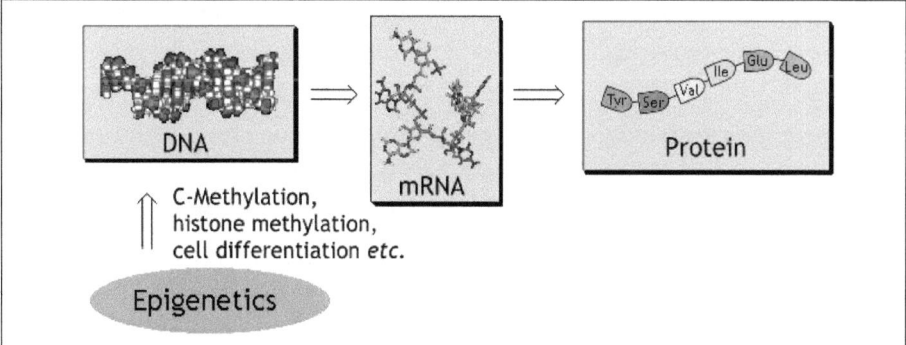

Figure 1: Function of DNA and regulation of gene expression by epigenetics

In order to understand the complex coherence between a gene and the phenotype of an individual (i. e. expression of proteins), the sequence of the DNA is of high interest. Hence several motivated research groups investigated in sequencing of biological relevant genes: The first innovative breakthrough was achieved by Maxam and Gilbert as well as by the Sanger group in the 1970's. Both research groups published the first two effective and competitive techniques for the determination of a DNA sequence[2,3].

The Maxam-Gilbert method[2] is based on chemical degradation of the DNA template, which means that the four nucleobases are cleaved in nucleobase-specific reactions, giving DNA fragments of different sizes. The lengths of these ^{32}P-radiolabeled fragments identify the positions of the nucleobases in the sequence. These fragments from four base-specific reactions are resolved by their size by polyacrylamide gel electrophoresis (PAGE) and the sequence can be directly read out from the band pattern provided by autoradiography. The accuracy of this sequencing method is limited by the length of the DNA template: The method is applicable for sequences not longer than ca. 750 nucleobases (duration of sequencing roughly one day). One big success in the application of the Maxam-Gilbert method was the decoding of the DNA sequence of bacteriophage T7 in 1983[4].

The competitive technique developed by Sanger et al. is an enzymatic sequencing method which was published in the same year as the Maxam-Gilbert sequencing (1977). Similar to the "plus and minus" method[5], the single-stranded DNA-template is extended by the addition of 2',3'-dideoxynucleotides which act as specific chain-terminators for the DNA polymerase. The resulting radiolabeled fragments from the PCR products are separated and detected as described for the Maxam-Gilbert method (PAGE). But in contrast to the chemical sequencing technique, the Sanger method delivers the sequence of the complementary strand of the DNA template. Translation of the determined sequence into its opposite gives the requested one. With the Sanger technique in hand the decoding of the bacteriophage ɸX174 sequence in less than one year consisting of roughly 5000 nucleobases was achieved[6] (Nobel Prize in 1980). Unlike the Maxam-Gilbert method, the Sanger technique was refined and commercialized which led to its broad dissemination throughout science and clinical diagnostics. The Sanger method has a few advantages over the Maxam-Gilbert technique. The reason for this is that it is a faster and more accurate technique, enabling the decoding of longer DNA templates in shorter time (standard template of 1000 nucleobases sequenced in 1-2 hours). Second, the Sanger method enables alternative labeling techniques like fluorescence labeling that avoids the use of toxic radioactive ^{32}P-labeled material.

Third, the Sanger method can be automated by use of fluorescence detectors and is therefore applicable to the routine laboratory (i. e. dye-primer[7] or dye-terminator[8] sequencing using capillary electrophoresis). The biggest milestones in the application of Sanger sequencing were the determination of the whole genome of *E.coli* bacteria[9] (4,639,221 bp) and finally the decoding of the whole human genome (3,069,431,456 bp), published by Craig Venter in Science 2001[10]. In 2003, this goal was accomplished by the Human Genome Project (HUGO) in a 13-year effort with an estimated cost of $2.7 billion. As a comparison to this, it was possible in 2008 to sequence the human genome over a 5-month period for ca. $1.5 million[11].

These overwhelming results encouraged the development of new and improved sequencing strategies, since the DNA-sequence information from humans and microorganisms has the potential to improve many facets of human life and society, including the understanding, diagnosis, treatment and prevention of diseases. During the past 5 years, "next-generation" sequencing technologies have rapidly evolved opening new perspectives in diagnostics.

1.1.2 The need for second-generation sequencing

Although the whole human genome has already been deciphered and also more than 150 different genomes are sequenced and partially analyzed, there is still a strong need for detailed sequence information in order to understand the functions of many genes. Such information would benefit the development of personalized medicine and significantly improve human health and the quality of life. The two already described sequencing methods, the enzymatical[3] and the chemical one[2] have in common that they use the time-demanding and expensive gel electrophoresis. Although the Sanger method used to be the method of choice within the last 30 years of sequencing, it is too laborious and expensive to meet future demands. This method is mainly based on fluorescence detection of dideoxynucleotide-terminated DNA extension products that are resolved by PAGE or capillary electrophoresis systems[12,13], where the sequence "read length" is no longer than approximately 1000 nucleotides. To decipher 3 billion bases more than 3 million runs would be needed, so that - together with overlaps and resequencing - the number of sequencing runs would be several millions.

For an improved insight into the structure and function of the human genome as well as for full realization of individual health care a need for much faster, simpler and cheaper sequencing technologies than the PAGE-based ones exists. Therefore several techniques were developed within the last two decades, such as pyrosequencing[14a-e], mass spectrometry-assisted sequencing[15a-c], sequencing by hybridization[16a-c], sequencing by ligation[17a,b], sequencing of single DNA molecules[18a,b] and sequencing by synthesis[19a-d], with some of them already having led to commercialized sequencers.

1.2 Sequencing by hybridization

The technique of Sequencing by hybridization (SBH) was first brought up in the late 1980's by Drmanac et al. introducing the read-out of a 100-bp sequence by a gelelectrophoresis-free method[16a,20]. This method enables the determination of a immobilized sequence by the maximal overlap and hybridization of its constituent octamers (or nonamers)[16a]. The 100-bp segment from 922-bp EcoRI-Bgl II human genomic fragment containing the β-interferon gene, which had already been sequenced[21], served as template for resequencing and for the proof-of-principle. Based on the evaluation of this method, Drmanac et al. successfully applied this technique on sequencing of larger templates such as the determination of a few sequence regions in the p53 gene[16c]. Also other research groups designed SBH-approaches for the determination of genomic sequences: Broude et al. developed the so-called termed position SBH (PSBH) published in 1993[16b] which is very similar to the method developed by Drmanac et al.

The PSBH method is based on the use of immobilized duplex probes that contain single-stranded 3'-overhangs instead of single-stranded probes (see Drmanac[16a]). The PSBH-principle is shown in *figure 2a* in detail: The duplex probe possesses a double-stranded constant region and a variable single-stranded 3'-overhang. The probe is 5'-biotinylated and immobilized on Streptavidin-coated magnetic beads, then treated with the single-stranded target DNA (^{32}P-labeled on 5'-terminus) annealing selectively to the 3'-terminal sequence of the duplex probe. The double-stranded probe provides sequence stringency in detection of the 3'-terminal sequence of the target DNA caused by base-stacking between the preformed DNA duplex and the newly formed duplex[22]. The DNA ligases' adjacency to the duplex probe-target complex is directed by the probe-target configuration, also the enzyme is susceptible to single-base mismatches near the ligation site[23]. As a consequence of this, the ligation step allows discrimination between perfectly matched and mismatched target DNAs.

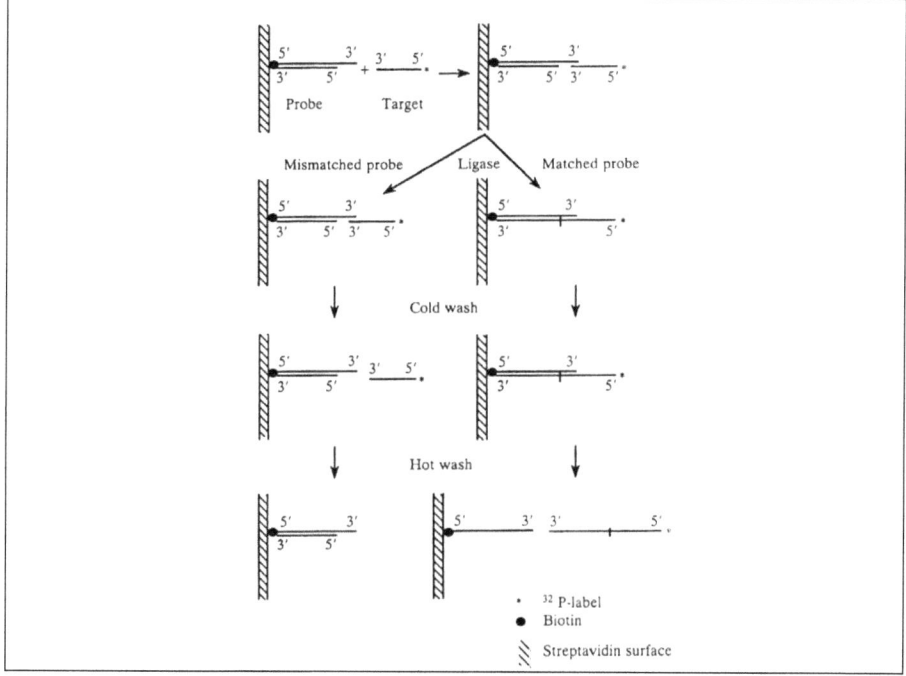

Figure 2a: The PSBH-method: Outline of experiments to test ligation discrimination for matched and mismatched targets bound to duplex probes[16b]

Both unligated and ligated targets are removed by washing the beads at 4 °C (cold wash) first, then at 90 °C (hot wash): At 90 °C the nonbiotinylated strand is melted from the immobilized biotinylated strand. The ligation efficiency is taken as the ratio of ^{32}P released at 90 °C to the total amount of ^{32}P used and defined as the ratio of ligated target to the total target used. The discrimination factor therefore is defined as the relative ligation efficiency of perfectly matched and mismatched targets. One attempt to enhance the discrimination between matched and mismatched target sequences was the polymerase extension reaction of the free 3'-terminus of the duplex probe (see figure 2b).

Extension of the duplex probe's free 3'-terminus with a ^{32}P-labeled dNTP is performed using a DNA-polymerase with lack of 3'->5' exonuclease and terminal transferase activity. This process requires a correct duplex formation between the free 3'-terminus and the ligated target to be able to distinguish between matched and mismatched probes.

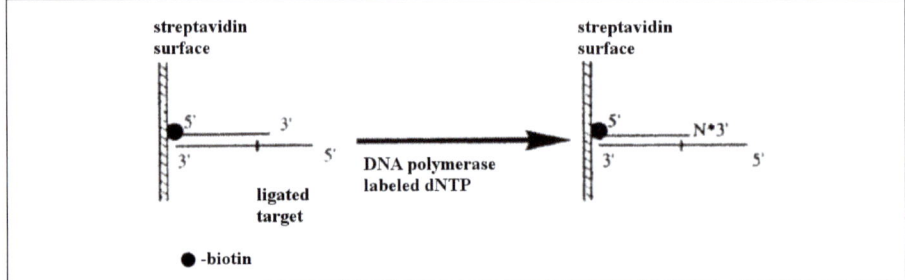

Figure 2b: PSBH enhancement by the use of polymerase extension reaction[16b]

Based on the disadvantage that the PSBH-method only detects 3'-terminal sequences, Broude et al. improved this technique by designing a set of nested 3'-deletions possessing a common 5'-terminus derived from the target sequence. The positional information about the distance between the 3'-sequence and a known reference point in the sequence was achieved by using such a kind of nested targets (i. e. overlapped fragments). One decade later, the nested target SBH method was defined by Shamir et al. as the reconstruction of a DNA sequence that is based on its k-mer content with k as length of the target fragments[24].

According to Drmanac[25] the SBH method has the advantage of reading whole "words" and not only single-bases within a sequence, but this technique is hampered in broad application by its disadvantages like biochemical difficulties and a high error rate during the hybridization process for longer sequences (> 400 bp)[24]. Thus the need for improvement of the read-out of longer and unknown sequences by complex algorithms[24,26] still remains challenging. The SBH technique is also hardly applicable to de novo-sequencing, although it is a reliable method for resequencing.

1.3   Pyrosequencing

Pyrosequencing is a nonelectrophoretic, real-time DNA sequencing method which was first reported by Hyman et al. in 1988[27]. The sequencing technique is based on the detection of released pyrophosphate (PP_i) during the DNA polymerization reaction. Similar to the Sanger method, the complementary strand is synthesized via PCR and can be translated afterwards into its opposite. Ronaghi et al. significantly improved the pyrosequencing technique and thus published two pyrosequencing methods in the 1990's, the solid-phase[14b] (see *figure 3a*) and the liquid-phase method[14c] (see *figure 3b*).

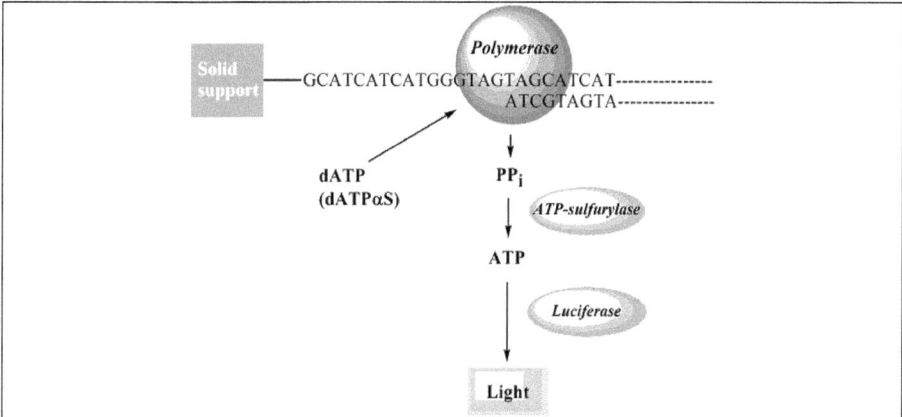

Figure 3a: Scheme of solid-phase pyrosequencing according to Ronaghi et al.[14d]

Solid-phase pyrosequencing is enabled by the immobilization of the DNA-template on an array (e. g. Streptavidin-coated magnetic beads) and treatment with the dNTPs and a three-enzyme system: The visible light which is generated during a cascade of enzymatic reactions is proportional to the number of incorporated nucleotides. These coupled enzymatic reactions were already successfully applied by Nyren et al.[28] for certain polymerase activity assays.

As depicted in scheme in *figure 3a*, the nucleotide incorporation done by the polymerase (often the Klenow fragment of *Escherichia coli* DNA Pol I) leads to the release of inorganic PP_i, which is then converted by ATP sulfurylase[14d] into ATP. The released ATP is an energy source for the enzyme luciferase (from the American firefly *Photinus pyralis*), which subsequently oxidizes luciferin and consequently generates light with the wavelength of 560 nm. The amount of emitted light is detected by a charge-coupled device (CCD) camera or a photodiode. The whole process, from DNA polymerization to light emission, happens within 3-4 seconds. The concentration of DNA polymerase is selected to be much higher than the template concentration enabling the immediate start of polymerization reaction. For each incorporation step, one particular known nucleobase (dATP, dCTP, dGTP or dTTP) is added and the light signal can only be detected if the appropriate nucleotide is incorporated complementary to the template strand.

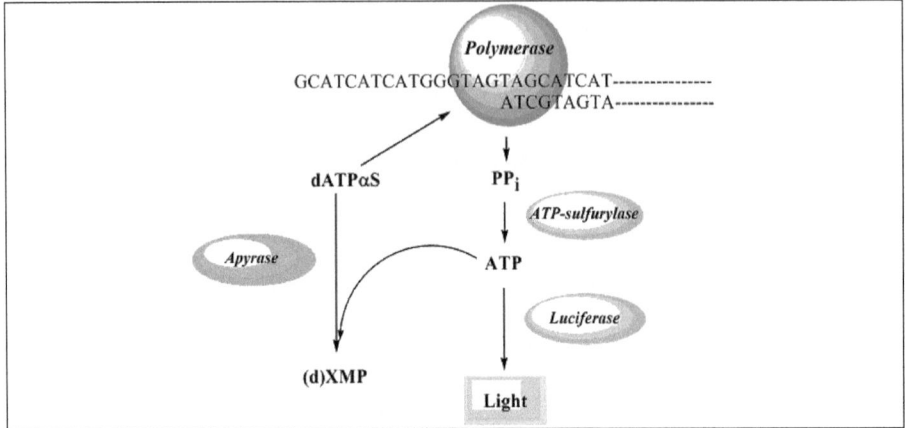

Figure 3b: Scheme of liquid-phase pyrosequencing according to Ronaghi et al.[14d]

After light signal detection, the array is washed for removal of excess substrate and the next incorporation cycle can take place. Ronaghi et al. often observed false signals when dATP was added to the solid-phase sequencing system. The reason for this is that dATP is not only a substrate for the incorporation by the polymerase, but also for the luciferase. The problem of competitive reactions resulting in false signals could be solved by substitution of dATP with dATPαS[14b]. The sulfur-containing nucleotide was found to be efficiently incorporated into the template while being inert towards the luciferase. A solid-support free method of pyrosequencing, named liquid-phase pyrosequencing, is enabled by the extension of the three-enzyme system to a four-enzyme system introducing the nucleotide degrading enzyme apyrase (see *figure 3b*). With the use of apyrase no intermediate washing step is required so that pyrosequencing can take place in solution. Apyrase possesses a high catalytic activity and therefore degrades the unincorporated dNTPs (products marked as (d)XMP in *figure 3b*) by hydrolyzing the triphosphates first to form diphosphates and finally monophosphates.

This four-enzyme system combined with the removal of inhibitory substances and with minimization of the dilution effect enables accurate sequencing of 200 nucleobases. The signal-to-noise ratio of pyrosequencing is relatively high, but longer reads are limited by misincorporation. Misincorporation of a dNTP terminates the primer strand causing reduction of the signal intensity. This problem is even more inherent to *de novo*-sequencing of polymorphic regions in heterozygous DNA templates.

However, pyrosequencing allows high-throughput sequencing and is a powerful technique for genotyping (*i. e.* detection of SNPs), microbial typing, resequencing, tag-sequencing and for analysis of difficult secondary structures such as hairpins.

The pyrosequencing technique is also applicable to different instrumentation methods like solid-phase pyrosequencing in microfluidics[14e] or pyrosequencing in an automated microtiter plate format[29].

The first next-generation DNA sequencer based on the pyrosequencing technique was released to the market in 2005 which was developed by the 454 Life Sciences Roche company. With this instrument (named Genome Sequencer FLX with GS FLX Titanium series reagents™) it is possible to sequence approximately 400-600 megabases of DNA per 10-hour run[11]. The technology is known for its straight sample preparation and long sequence reads (400-500 base pairs in length) with high accuracy at the same time (including paired reads).

Here, the already described solid-phase pyrosequencing method is embedded into a cyclic array sequencing method[30] as shown in *figure 3c*: The adapter-ligated DNA fragments are immobilized on small DNA-capture 28µm-beads in an emulsion and clonally amplified by PCR. These DNA-bound beads are randomly placed into a ~29 µm well (~29 µm is the dimension where only one single bead fits per well) on a PicoTiterPlate then and treated with a mix of DNA polymerase, ATP sulfurylase, and luciferase. The whole loaded PicoTiterPlate is placed into the sequencer, where one surface of this semi-ordered array has the function of a flow cell allowing reagent addition and removal, whereas the other surface enables CCD-based signal detection while it is bound to a fiber-optic bundle[30]. The PicoTiterPlate is treated with the sequencing reagents (containing buffers, one particular nucleotide per cycle (daTPαS, dCTP, dGTP, dTTP) and luciferin) across the wells. During a sequencing run the four dNTPs are added sequentially in a fixed order across the PicoTiterPlate device enabling massive parallel sequencing. Pyrosequencing and therefore light emission only takes place in wells were the particularly added nucleotide is incorporated into the bead-supported template and the released light signals are then recorded by the CCD camera in the instrument. The signal strength is proportional to the number of nucleotides incorporated: homopolymeric sequences generate a stronger light signal than single nucleotides during one nucleotide flow.

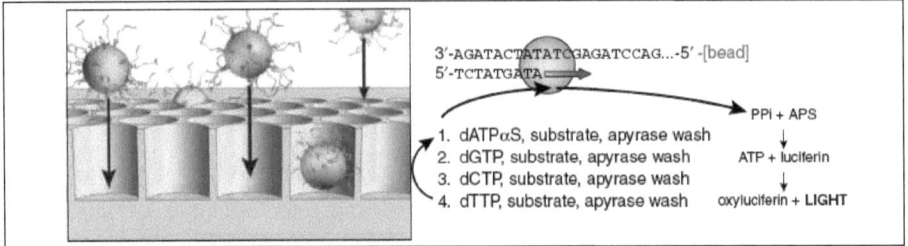

Figure 3c: Cyclic array pyrosequencing method applied to Roche's 454 sequencing system[29,30]

A major limitation of the 454 technology as well as of pyrosequencing in general is the appearance of false signals during sequencing of homopolymeric regions (such as AAA or GGG). Multiple incorporation of one particular nucleobase per one nucleotide flow cycle cannot be prevented by any terminating function which makes the interpretation of the detected light signal intensity inaccurate. Nevertheless, one big advantage of the 454 system with optimized pyrosequencing technique is its high read length: With the use of the 454 sequencer deciphering of the complete Neanderthal mitochondrial genome was achieved thus leading to a big publication[31] in *Nature* in 2006.

1.4   Sequencing by synthesis

1.4.1   The invention of reversible terminators for SBS

One of the first prototypes of the PAGE-free method called Sequencing-by-Synthesis was published by Metzker *et al.* named *"Base Addition Sequencing Scheme"* (BASS)[32]. Like Sanger sequencing, this method is an enzymatical technique but uses 3'-O-modified dNTPs instead of ddNTPs as chain terminators. These nucleotides have the ability to stop the polymerase reversibly after the incorporation of one modified nucleotide. The scheme of the BASS method is shown in *figure 4a*. The DNA template is biotinylated and bound to a solid support. After primer annealing, the four 3'-blocked dNTPs are added, each of them possessing a spectroscopically unique blocking group.

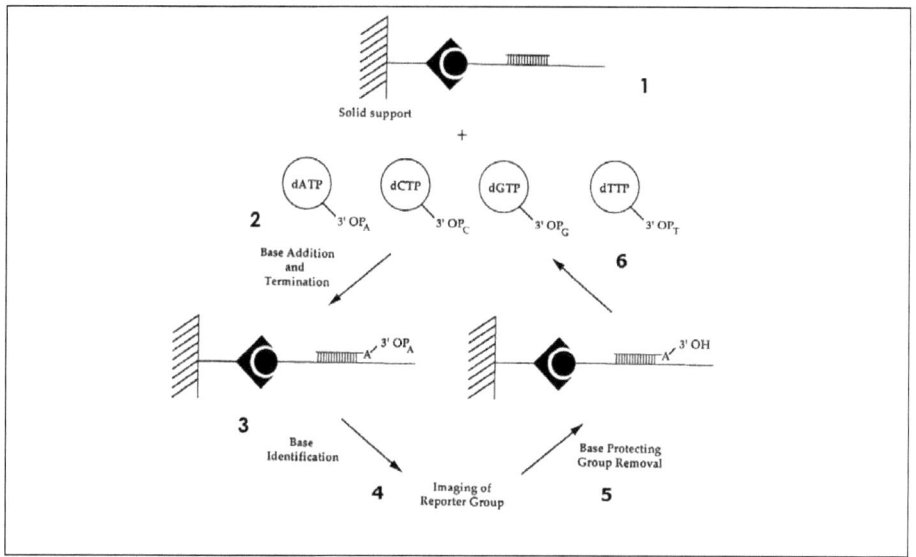

Figure 4a: Principle of the Base Addition Sequencing Scheme (BASS)[32]

Several polymerases were found to accept and incorporate these 3'-modified nucleotides, such as the Klenow fragment of DNA polymerase I, AmpliTaq® DNA polymerase, Vent$_R$® (exo⁻) DNA polymerase and others[32]. The polymerase is immediately stopped after the incorporation of one single 3'-modified dNTP and the base-specific 3'-blocking tag is detected and photolytically cleaved. With the 3'-group removal, the 3'-end of the primer is regenerated and the next incorporation-detection cycle is enabled. Because of their 3'-blocking group stability during polymerase incorporation, but photolytical cleavability of this 3'-tag, these 3'-modified nucleotides were named "reversible terminators".

The publication of Metzker introducing the BASS method presents only one successful cycle of incorporation, detection and 3'-tag cleavage. Seven different fluorescent aromatic labels were used as 3'-modifications, which were tolerated by the polymerase differently among each other: It could be determined that 3'-O-(2-nitrobenzyl)-2'-deoxyadenosine was incorporated best compared to the other 3'-blocked nucleotides which weren't employed for a whole BASS cycle.

Over a decade later, Metzker et al. surprisingly found out that the claimed structure of the reversible terminator, 3'-O-(2-nitrobenzyl)-2'-deoxyadenosine, was incorrect and that the correct structure of this terminator was assigned as N^6,N^6-bis-(2-nitrobenzyl)-2'-

deoxyadenosine[33] by NMR spectroscopy. The confusion about the structure of 2'-nitrobenzyl-labeled dATP motivated the Metzker group to synthesize both compounds as well as the monobenzylated nucleotide N^6-2-nitrobenzyl)-2'-deoxyadenosine and compare their acceptance and termination properties in polymerase assays. As a result of these tests, the researchers established that the 3'-blocked dATP acted as a poor substrate for the polymerase compared to the other both N^6-mono- and dialkylated dATPs[33]. The latter ones showed good incorporation properties and were regarded as promising candidates for further SBS applications.

Until now, the BASS method - later named as cyclic reversible termination (CRT) - was not developed further, but due to the presentation of these labeled dNTPs the concept of a "reversible" terminator was born.

Nearly at the same time a 3'-O-dye-labeled terminator was synthesized and published by Hovinen et al.[34a] in 1994, which can be regarded as one of the first complete terminators (see *figure 4b*).

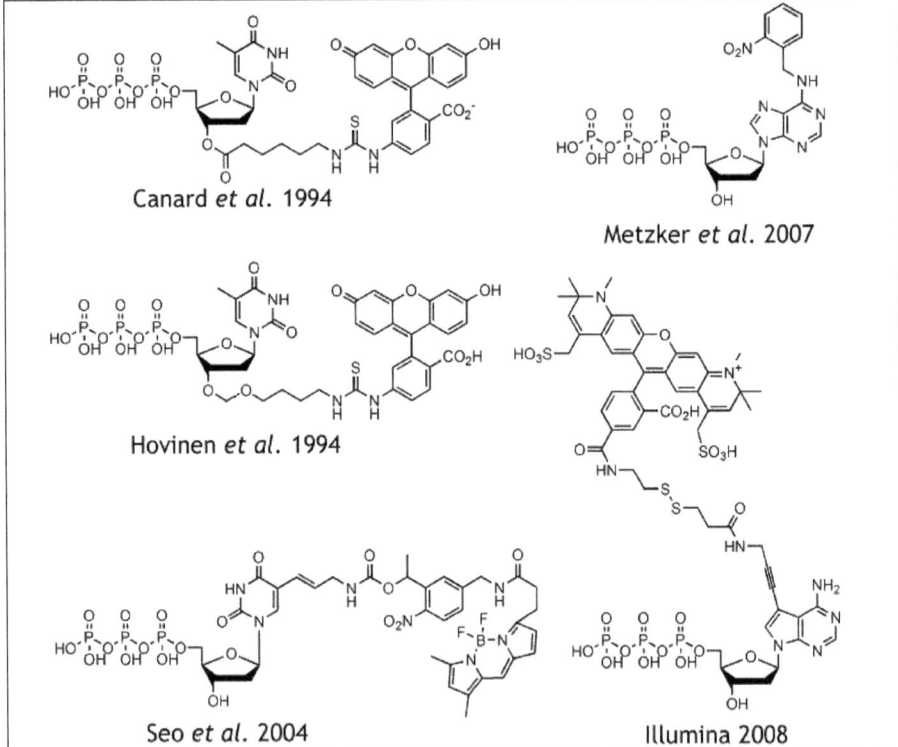

Figure 4b: Variety of reversible terminators for different SBS approaches

Compared to Canard's[34b] and Metzker's[33] terminators, this dTTP derivative possesses an aminoalkoxymethyl linker which is attached to the 3'-hydroxy function and labeled with fluores-ceinisothiocyanate (FITC). This labeled nucleotide shows good termination properties by using the thermostable Tet/z-DNA Polymerase, an enzyme that accepts and incorporates this substrate in high fidelity.

One significant difference between Canard's and Metzker's terminator is the cleavability of their particular 3'-modification. Hovinen's nucleotide is supposed to carry a remarkably stable 3'-dye-linker system that may not be cleavable under mild conditions. Therefore this nucleotide cannot be regarded as reversible terminator, although it has terminating properties. These 3'-modified reversible terminators developed in the 1990's were used for the first attempts in developing the SBS method, but one decade later, another type of reversible terminators became popular: Seo et al. introduced in 2004 new 3'-unmodified nucleotides wearing a fluorescent label attached to the base motif as reversible terminators, enabling a couple of mass spectrometry-assisted SBS cycles[35a,b]. Companies such as Solexa (now Illumina) and Roche came up with publications[19d,36] and patents[37,38a,38b] covering the design and synthesis of reversible terminators, which dye-linker system is annealed to the nucleobase moiety, accepted by the polymerase and quantitatively cleavable (see *figure 4*). This clarifies that the 3'-blocking group is not absolutely necessary for any SBS technique which also depends on some other issues.

In most of the cases, the reversible terminators presented have to possess the following properties for fulfillment of a PCR-based SBS method:
- The nucleotide is a 5'-triphosphate.
- The nucleotide is labeled with a spectroscopically unique group like a fluorophore or dye, attached either to the base moiety or to the 3'-end.
- The modified nucleotide is accepted and incorporated into the template by the polymerase.
- The modification (3'-blocking group, fluorophore, linker etc.) is

quantitatively removable after incorporation and detection of the nucleotide.

Nucleotides addressing all these points would have the potential to enable the development of a Sequencing-by-synthesis technique leading to high-troughput DNA sequencing.

1.4.2 The array-based SBS technology

The array-based sequencing-by-synthesis(SBS)-technology is a promising diagnostic tool which provides fast and cost-effective sequence information about point mutations in genes of short read-length. For the determination of short gene sequences, no time-consuming gel electrophoresis is needed because the sequence can be directly read out by a four-color code. Within the last years, several researchers from the industrial and academic field have focussed on the development of an accurate SBS-technique using 3'-O-modified reversible terminators, which are efficiently incorporated into the DNA-template by a highly-tolerable polymerase.

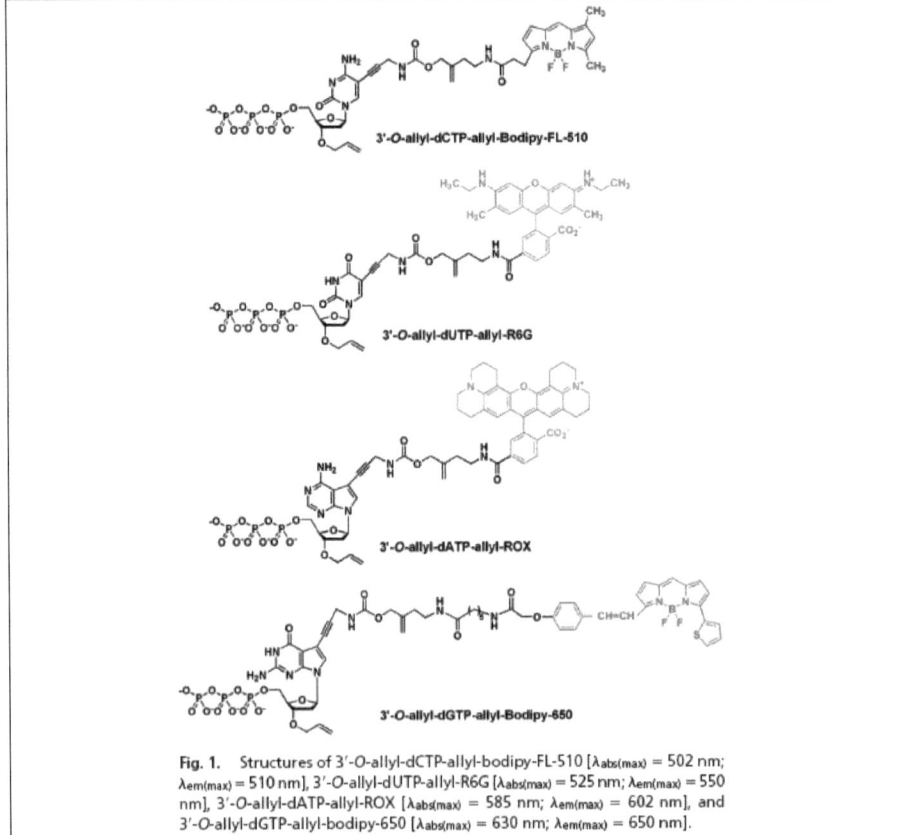

Fig. 1. Structures of 3'-O-allyl-dCTP-allyl-bodipy-FL-510 [$\lambda_{abs(max)}$ = 502 nm; $\lambda_{em(max)}$ = 510 nm], 3'-O-allyl-dUTP-allyl-R6G [$\lambda_{abs(max)}$ = 525 nm; $\lambda_{em(max)}$ = 550 nm], 3'-O-allyl-dATP-allyl-ROX [$\lambda_{abs(max)}$ = 585 nm; $\lambda_{em(max)}$ = 602 nm], and 3'-O-allyl-dGTP-allyl-bodipy-650 [$\lambda_{abs(max)}$ = 630 nm; $\lambda_{em(max)}$ = 650 nm].

Figure 5: Structures of the four reversible terminators for Ju's SBS method[19b]

The invention of new sequencing technologies

Especially the Columbia University research group developed a SBS technique based on reversible terminators possessing a 3'-O-modification[19b,39a-d], e.g. an allyl-group[19b,39c-d] and a dye-labeled cleavable linker containing a particular fluorophore for each nucleobase (see *figure 5*): The four nucleobases carry four different dye-labels attached to the base motif with a cleavable linker that contains an allyl moiety. An allyl function serves as 3'-blocking group which is elegantly cleavable under the same conditions like the dye-linker system: In aqueous buffered solution, the nucleotide is deallylated quantitatively by incubation of the template for 5 min at 60 °C using a Pd-catalyst. The four reversible terminators employed for the SBS approach are illustrated in *figure 5*. The principle of the SBS method is shown in *figure 6a part A*:

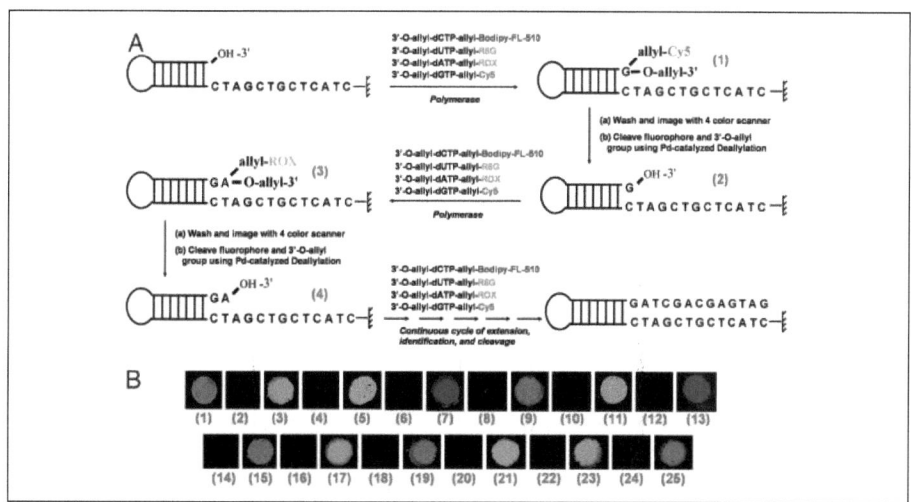

Figure 6a: Scheme of the array-based SBS method according to Ju et al.[19b]

Step 1: A hairpin-shaped self-priming template immobilized on a chip is treated with all four reversible terminators (dNTPs with N = A, C, G, T and different dye label on each nucleobase) and a highly tolerable polymerase, which incorporates the correct nucleobase complementary to the template strand (at 68 °C within 10 min incubation time).

Step 2: After one washing step and the capping reaction (*i. e.* blocking of unreacted primers with unlabeled 3'-O-allyl-dNTPs), the fluorescence signal is detected and the dye-linker system as well as the 3'-blocking group is cleaved off using Thermopol I reaction buffer/Na_2PdCl_4/$P(PhSO_3Na)_3$ and incubation for 5 min at 60 °C.

Step 3: The terminal 3'-group is regenerated, the chip immersed in a 3 M tris·HCl buffer (pH 8.5) and incubated again for 5 min at 60 °C for Pd-cat removal. The surface is washed again and scanned for confirmation of complete fluorophore removal. The second elongation cycle is enabled and a new mixture of polymerase and the four labeled reversible terminators is added.

The four-color fluorescence scanner delivers a plot of the raw fluorescence emission intensity at the four designated emission wavelength of the four chemically cleavable reversible terminators (see *figure 6b*).

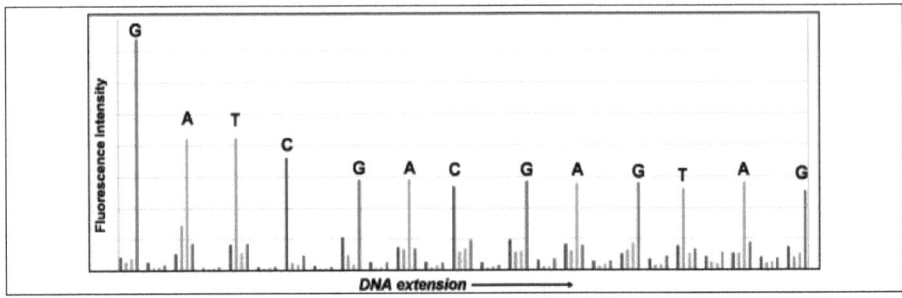

Figure 6b: Four-color sequencing data plot obtained from SBS[19b]

The sequence of the template can directly be read out by its color code obtained from the raw data plot (see *figure 6a part B*) without any processing.

Before the fluorescence scanner was used as detector, the SBS method has been evaluated in solution and each incorporation and deallylation step was checked by MALDI-TOF analysis[19b,39c-d]. With this first SBS experiment the efficiency of the reversible terminators regarding their incorporation and cleavage was confirmed. The SBS system was transferred to an array system then[19b]: As the DNA chip offers a large surface, many templates can be sequenced in parallel leading to simultaneous sequencing of a large number of DNA templates.

When Ju et al. launched their first *de novo* sequencing on the chip, they had chosen a self-priming DNA template which was treated with a solution containing all four dye-labeled reversible terminators and the 9°N mutant DNA polymerase. One important thing to mention here is that a so-called synchronizing step (also known as capping step) had to be introduced for avoiding any lagging fluorescent signal caused by previously unextended priming strands[19b]. One big advantage over pyrosequencing as claimed by Ju et al. is the accurate determination of homopolymeric sequences by using the SBS method, a demand that is only poorly fulfilled by pyrosequencing. In order to confirm this fact, Ju et al. made a sequencing experiment[19b] with a self-priming DNA template bearing two homopolymeric regions (10 T's and 5 T's) and used their SBS approach as well as pyrosequencing for comparison. As a result of this, all bases as well as the homopolymeric sequences of the template were unambiguously identified by the SBS method. Pyrosequencing of the same template gave an unprecise sequence data: The first four individual nucleobases were identified, but the homopolymeric regions caused two big broad peaks in the plot not exactly quantifying the number of the homopolymer. With the SBS technique in hand Ju et al. could sequence up to 20 nucleobases with high accuracy, making this technique interesting for automatized application.

By 2006, the Illumina (former Solexa) Genome Analyzer as a "short read" sequencing platform based on the SBS technique was commercially launched. The illustration in *figure 6c* displays the sequencing principle:

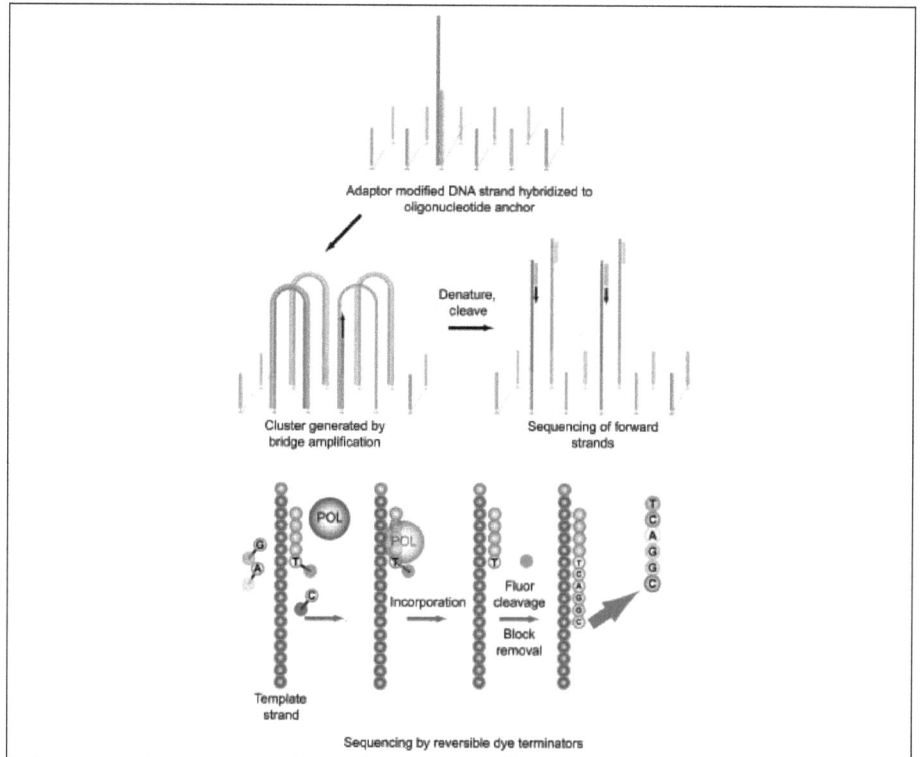

Figure 6c: Scheme of the Illumina Genome Analyzer sequencing[11]

In a flow cell consisting of an optically transparent slide where oligo-nucleotide anchors are bound, an adapter-modified, single-stranded DNA template is added and hybridized to these anchors. The bound templates are amplified then by "bridged" PCR: The captured DNA strands bend over and hybridize to an adjacent anchor. These arched templates are then amplified with PCR converting the single-molecule DNA template into a clonally amplified arching "cluster". The clusters are denaturated, chemically cleaved and washed in a manner that only the single-stranded templates remain on the surface. The primer is then hydridized complementary to the adapter sequences and a mixture of polymerase and four differently dye-labeled reversible terminators are added. The polymerase incorporates the dye-labeled nucleotide complementary to each template strand in a clonal cluster, then the excess of reagents is removed by a washing step and the clusters are optically scanned by recording the fluorescence signals. The reversible dye terminators are then chemically unblocked, the fluorescence labels are removed and washed away and the next sequencing cycle can take place. With read-lenghts of 36 bases within 2.5 days the Illumina Genome Analyzer sequences much slower than the 454 pyrosequencing system from Roche.

One disadvantage of the Illumina sequencer is that the base-call accuracy is lowered by sequencing longer read-lengths[11]. This may be caused by under- or overincorporation of the nucleotides, also the removal of the blocking group might sometimes fail. These signal aberrations accumulate during successive sequencing cycles and form a heterogeneous population within a cluster. As a result of this, the fluorescence signals purity decreases and the base calling is hampered. Investigations in improving the sequencing technique as well as the development of algorithms for data-image analysis and interpretation are in progress.

1.4.3 The EU-project "ArraySBS" and its aim

The SBS method offers a new technology for detecting point mutations (SNPs) or homopolymeric regions very fast and in high accuracy. Within the last 5 years, the invention of a commercially available sequencer using the SBS technique based on 3'-modified dye-labelled dNTPs was a big leap forward for the development of new sequencing techniques. Although the SBS technique is currently only applicable to shorter DNA sequences, it has already been put into practice and can be improved.

As a consequence of this, competitive projects in developing an applicable SBS technique were started in the last years, such as the EU craft project "ArraySBS". It was funded by the EU and started in august 2005 consisting of a scientific and industrial consortium. The aim of the ArraySBS project was to provide proof-of-principle for the SBS setting on an array of primer features, where the 3'-end of the primer is extended with a reversible terminator. The project and its participants were divided into five main objectives, each of them dealing with different particular operational objectives as follows:

- The development and synthesis of four dNTPs (A, C, G, T) with blocked 3'-end and labeled with a fluorescent dye, *i. e.* the four complete reversible terminators.

- The identification, isolation and development of a DNA polymerase that accepts these four dye-labeled reversible terminators including the polymerases' evolution and improvement.

- The development of a microfluidic device prototype that allows array SBS reactions by integrating incubation steps at different temperatures with controlled agitation of reaction components over the entire area of a microarray, including washing and regeneration procedures.

- The development of a software prototype for primer design and for analysis of sequencing-by-synthesis generated data. Besides this objective, the base-caller should be modified and suitable primers for the *p53*-resequencing should be identified. As an outcome of this, software for primer prediction and sequence data handling should be delivered at the end of the project.

- The validation of the SBS technology through the resequencing assay of a coding sequence of a gene. At the end of the project, the array SBS procedure should have reached a proof-of-principle stage with up to 100 primer features on a microarray, in the best case the *p53* gene should be sequenced by using the array SBS procedure.

Besides the design and the development of 3'-modified reversible terminators, the identification of a DNA polymerase that is capable to accept these reversible terminators was considered as one of the central goals of the Array SBS project. Rather limited success of such experiments described in literature indicated it to be a highly challenging endeavor. Several aspects of the system designed to incorporate reversible terminators appeared to be critical. A polymerase suitable for our SBS-approach needs to possess proper affinity as well as selectivity towards modified nucleotides along with an adequate turnover rate and propensity to form terminated DNA which remains stable until the removal of 3'-modification by chemical means. In turn, the reversible terminator itself must display an appropriate balance between its chemical stability during enzymatic incorporation and the ability to be unblocked under conditions that are mild enough for the DNA structure to be maintained for further manipulations. The necessity for tight compatibility of both partners makes the challenge even more profound.

A strong collaboration within three years between the consortium members of the ArraySBS project should lead to the SBS proof-of-principle, putting great emphasis on the chemical and biochemical part of the project.

2 Development of an array-based SBS method

2.1 Selection of an appropriate polymerase for SBS

2.1.1 Function and properties of a polymerase

In any organism the replication of DNA happens based on a complex biological process that is supported by auxiliary enzymes such as ligases and helicases, but in principle catalyzed by polymerases. The first polymerase isolated was *E. coli* DNA Polymerase I obtained from cell extraction and purification of *E. coli* bacteria. This enzyme is one of the few well-explored polymerases due to the fact that its crystal structure is known[40]. The *E. coli* DNA Polymerase I has multifunctional properties within one polypeptide chain that contains various enzymatic activities which can be separated into different domains[41] forming active sites. Consequently, *E. coli* DNA Polymerase I does not only act as replicant agent for DNA, it has also the ability of repairing DNA.

For efficient PCR as well as for other applications such as DNA labeling, sequencing and amplification, a polymerase has to be reliable, accurate and fast. The property of incorporating the correct nucleobase into the template strand is called "accuracy" or "fidelity" of a polymerase. The fidelity of polymerases is still of high interest and until now not fully understood: How does the enzyme know which nucleotide is to be incorporated in one particular position complementary to the template strand?

First of all, recognition of the correct nucleotide by the polymerase is hardly ever caused by suitable hydrogen bonding between the complementary substrate nucleobase and the template nucleobase. Thus hydrogen bonding has a very low effect on the selectivity of the polymerase as it was demonstrated in studies by Kool *et al.*[42a,b] on base-pairing behavior of universal bases and natural bases. The geometric selection of a nucleobase given by its size and shape influences more strongly the incorporation accuracy.

Due to the sterical and electrostatical effects the active site of the polymerase accepts only certain geometrical conformations of both primer-template and nucleobase substrate[42,43]. An example of such a ternary complex, consisting of a nucleotide (ddCTP) and the template-primer system bound to the active site of rat DNA Polymerase β was published by Pelletier et al. in 1994[43]. Detailed data of this complex provided us with a deeper insight into the active site of a polymerase. The idea behind these studies was to crystallize polymerase β under nearly physiological conditions with and without complexing the template-primer-ddCTP system for spectroscopical analysis. From crystallization a sort of "freezed" complexes were obtained and resolved by X-ray diffraction analysis. Moreover Pelletier et al. utilized the advantage of stopping the DNA extension directly after nucleotide incorporation by the use of ddCTP instead of the naturally occurring dNTPs as substrates. One of the observations derived from the structural analytics was that the interaction of the enzymes' amino acid residues between the atoms of the nucleobases (from primer-template-complex) plays an important role in bringing the template into the active site and giving it a preferred conformation.

Another conclusion obtained from the X-ray structural analysis was that Mg^{2+} ions in the active site of the polymerase directly induce the nucleotidyl transfer reaction[43,44] (shown in *figure 7*): A new phosphordiester linkage is formed while the 3' hydroxyl of the terminal dNMP on the primer strand attacks the 5'-α-phosphate.

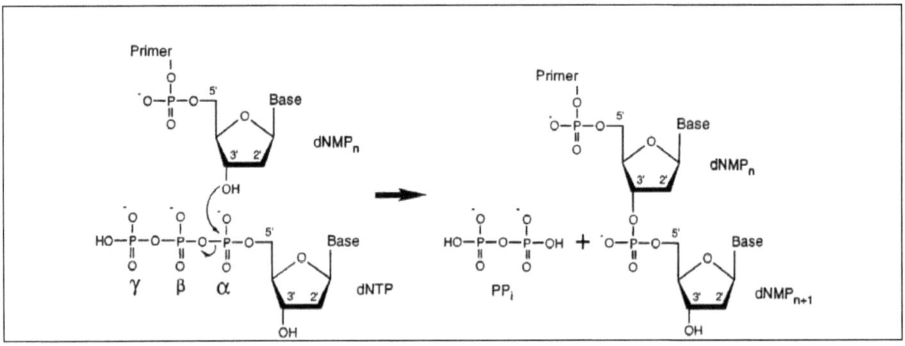

Figure 7: The nucleotidyl transfer reaction occurring in the active site of the polymerase[43]

In this manner the primer strand with n dNMPs is elongated up to (n+1) dNTPs then, resulting in the release of pyrophosphate (PP_i) and carrying a new free 3'-hydroxy function, where the next incorporation step can take place.

A scheme of the active site of the polymerase with its ddCTP binding pocket is shown in *figure 8* illustrating the transition state with the pentacoordinated α phosphate of ddCTP[43].

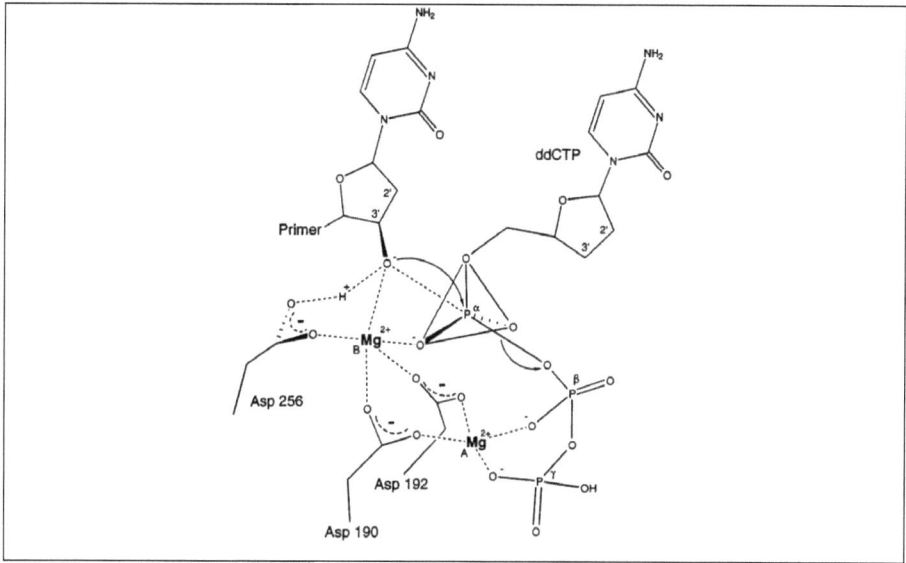

Figure 8: The transition state during ddCTP incorporation on the polymerase β active site[43]

In the binding pocket of the active site, two Mg^{2+} ions are clustered around the amino acids Asp 190, Asp 192 and Asp 256 from the polymerases' primary structure. One of these cations is bound specifically to the β and γ phosphates of ddCTP while the negative charge on the α phosphate is neutralized by the second cation and the attack of the 3'-hydroxy group can occur in an in-line fashion with release of pyrophosphate[43]. This means that during the transition state the attacking (3' hydroxyl of primer strand) and the leaving group (PP_i) have to be in linear position relatively to the α phosphate occupying the two apical positions of the pentacoordinated α phosphate. The polymerase β active site is very similar to the 3'->5'-exonuclease domain of E. coli polymerase I which also possesses a two-metal center involved in exonucleic activity[45]. It is nonetheless still not clear which catalytic role the particular amino acids of the enzyme's side chains from rat polymerase β and also of E. coli polymerase I play, although the structures of these enzymes are already resolved.

Polymerases that are optimal for DNA replication often possess the already mentioned 3'->5' exonuclease activity, which means in praxis that these enzymes do "proofreading"[43,45] during PCR reaction as depicted in *figure 9*.

Such a "high-fidelity" polymerase has the ability to recognize if mismatched pairing occurred during strand extension and moves the mismatch from the polymerization domain to the 3'->5'-exonuclease domain. At this site of the polymerase, the mismatched base is cut off and the DNA template strand is then transferred back into the polymerization domain. The proofreading effectiveness often depends on the sequence of the template: It was found that pyrimidine-rich sequences are more effectively proofread than purine-rich ones due to the lower stability of AT-stretches, a fact that facilitates the proofreading.

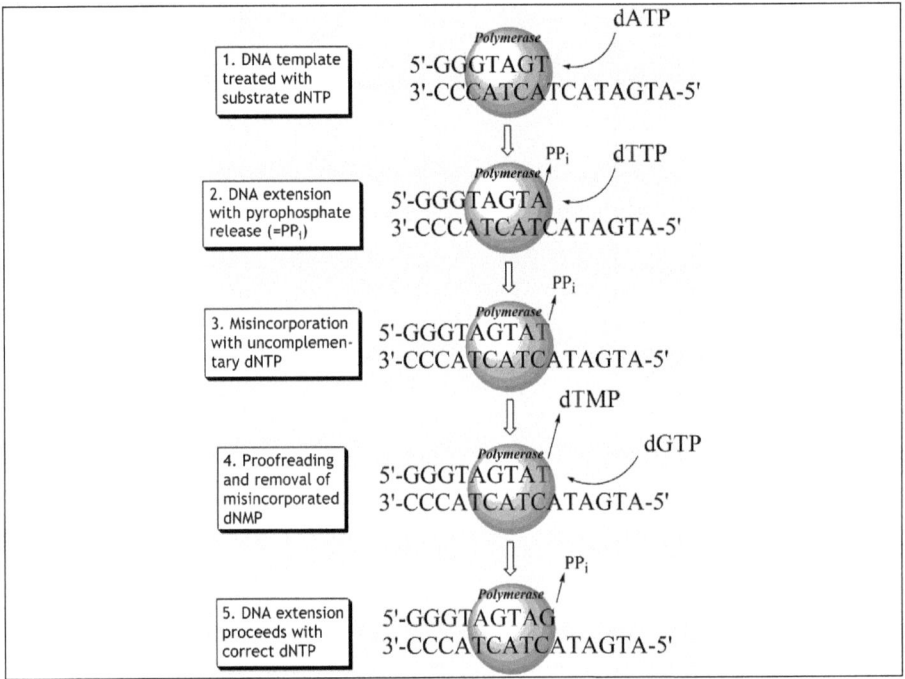

Figure 9: Proofreading (3'->5' exonuclease activity) during PCR

Based on the knowledge about the structure of polymerase's active site and the DNA polymerization behavior with ddCTP, Canard *et al.* found out that some polymerases lacking 3'->5'-exonuclease activity (such as *Taq* DNA Polymerase and Sequenase) are capable of incorporating 3'-esterified dNTPs into the DNA template[46]. Several incorporation tests were done using *Taq* DNA polymerase and high concentrations of 3'-modified dNTPs because natural nucleotides occurring as impurities were preferentially incorporated over the 3'-modified ones. The *Taq* polymerase showed to be tolerant against 2'-deoxy-3'-anthranyloyl nucleoside triphosphates (3'-ant-dNTPs) and incorporated these nucleotides correctly paired to the template strand. Canard et al. also made a fascinating observation when using especially Sequenase for the DNA extension: An 3'-esterified primer strand was synthesized chemically and hybridized with its complementary strand possessing an (dG)$_5$ overhang (see figure 10). This template (3'-ant-21-mer) was incubated with Sequenase and titrated with [α-^{32}P]dCTP, which means that this radioactively labeled substrate was added in various concentrations as indicated by lanes 1-9 of gel picture in figure 10 (e. g. lane 1: 2.5 nM, lane 4: 25 nM, lane 6: 50 nM and lane 9: 1000 nM).

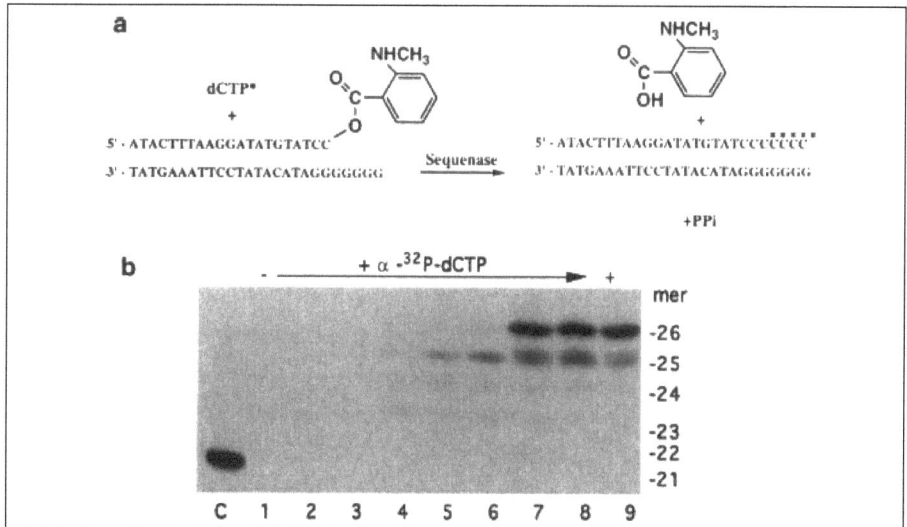

Figure 10: (a) Structure and deprotection reaction of 3'-ant-21-mer with Sequenase and ^{32}P-labeled dCTP; (b) Gel of the extension reaction under various dCTP concentrations (lane C: kinase-treated control)[46]

At a certain concentration of 3'-blocked dCTP, the Sequenase started DNA polymerization and the primer strand was extended up to a 25 or even 26 mer. This implies that the Sequenase has the ability of hydrolyzing the 3'-blocking group from the primer strand enabling incorporation and primer strand extension, a behavior which the Taq polymerase did not show during the incorporation tests. This unexpected result was named "catalytic editing", which means that Sequenase exhibits a strong 3'-esterase-like activity on the 3'-end of the DNA template[46]. In further incorporation assays done with Sequenase and 3'-amido dTTP and 3'-thioureido dTTP analogues[47] the enzyme showed the same result: It incorporates these 3'-blocked nucleotides and hydrolyzes them leaving a 3'-amino-terminated DNA chain. With these incorporation tests Canard et al. could demonstrate that nucleotides bearing a bulky blocking group on the 3'-position are accepted and incorporated into the template by several polymerases, some of them, e. g. Sequenase, even able to hydrolyze the 3'-blocking group.

The research group of Marx et al. investigated further in determining the criteria for both selectivity and fidelity of several polymerases like the Human DNA Polymerase and E.coli DNA Polymerase I[48]. The polymerase incorporation assays carried out with 4'-alkyl[49a] or 4'-acyl[49b]-modified 2'-deoxythymidine-5'-triphosphates stressed the strong dependence of the polymerases' selectivity from sterical effects. The close fitting of the Watson-Crick geometry to the active site of the polymerase is claimed as one very important factor for the selectivity in nucleotide insertion[48,49]. When a nucleotide is not properly inserted, polymerases often generate errors by deletion of a nucleotide or insertion of an additional nucleotide resulting in frame-shift mutations[48c]. As a reason for these errors misalignment of the primer template complex is claimed[48c]. This means that polymerases exhibiting high misalignment fidelity might form tighter binding pockets for the primer template binding. These enzymes would tolerate only little geometric deviation and therefore prevent frame-shift mutations resulting from nucleotide deletions and insertions. In contrast to that, low misalignment fidelity enzymes would tolerate more geometric deviations resulting in a decreased fidelity[48c].

In order to enhance fidelity and selectivity of DNA polymerases, combinatorial enzyme design[50] is an appropriate method for tailoring optimized polymerases for various applications. Combinatorial enzyme design produces libraries of DNA polymerase mutants which can be used for screenings of particular reaction conditions or substrates, e. g. the SBS technique.

2.1.2 Polymerase selection with unlabeled reversible terminators

The effectiveness of the sequencing-by-synthesis technique in principle depends on the quality of two features: The polymerases' fidelity and selectivity is crucial on the one hand, on the other hand there exists a strong need for suitable 3'-blocked nucleotides that are unambiguously accepted and incorporated by the polymerase. For the development of our SBS-approach, several principal sources of enzymes were considered for screening, including commercially available and Fermentas proprietary polymerases, representing all families of DNA polymerases. After an initial screening performed using a number of 3'-modified nucleotides, the list of polymerases was narrowed to 23 representatives with lower or even with absence of exonucleolytic activity.

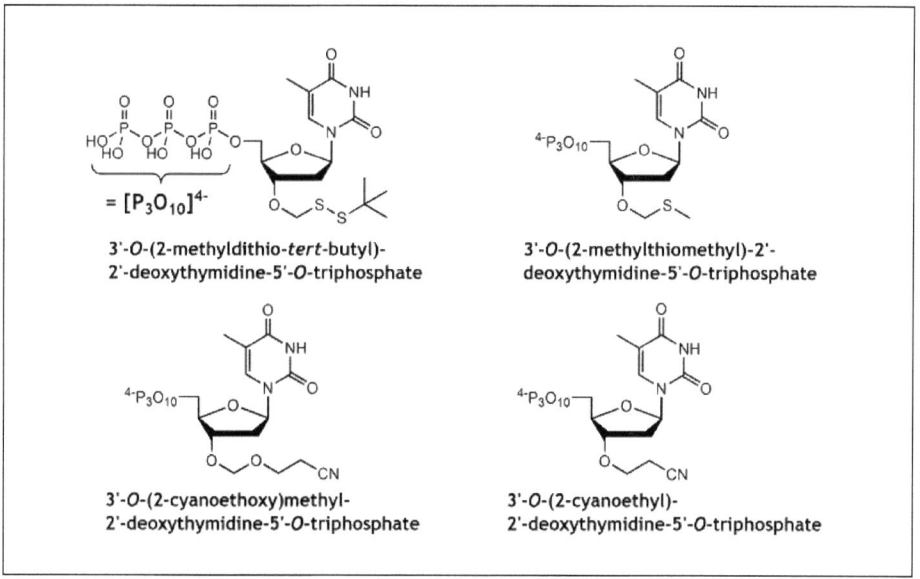

Figure 11: Our first 3'-modified reversible terminators without dye-labeling

Further polymerase acceptance tests were carried out with some selected unlabeled 3'-modified triphosphates shown in *figure 11*. For the SBS purpose only a few tolerated 3'-blocking groups are known in literature, which often are also patent-protected like e.g. the allyl function[51] (see *figure 5* in chapter 1.4.2). There was a need for alternative 3'-modifications that are not protected by patent like the already known and published 3'-blocking groups. A full-range screening program was executed focusing on the four unlabeled potential reversible terminators shown in *figure 11*, featuring methyldithio-*tert*-butyl (DTM), methylthiomethyl (MTM), (2-cyanoethoxy)methyl (CEM) and 2-cyanoethyl (CE) as 3'-protection groups albeit many more were investigated using the most promising polymerases for final evaluation, too. The DTM protective group as one of the first reversible 3'-blocking groups for chain terminators published by Kwiatkowski *et al.*[52] has been found to be unstable during the enzymatical incorporation into the DNA template. In contrast to that the screening performed using 3'-O-MTM-dTTP was highly successful. However, efforts in identifying an appropriate method for 3'-protecting group removal while keeping the DNA structure intact failed. As a consequence of this, we were looking for similar 3'-modifications such as the formacetal-type CEM and the ether-type CE protecting group. The 2'-deoxythymidine-5'-triphosphates possessing these groups as 3'-modifications were found to be incorporated very well into DNA by several polymerases and also terminated further primer extension. The results from the incorporation tests with these unlabeled 3'-modified triphosphates served as starting point for the design of the complete dye-labeled reversible terminators as well as for the polymerase evolution.

2.2 Polymerase acceptance tests

2.2.1 Materials and methods

The following incorporation assays were done at Fermentas' labs in Vilnius, Lithuania. An excessive amount of enzymes was used, at a range of ten times over the substrate DNA or more. In some cases the exonucleolytic activity of the polymerases was attempted to be inhibited by the addition of dTMP following recommendations found in literature[53]. The 3'-modified dTTPs were cleaned in a so-called "mop-up" reaction[54] for removal of the natural dTTPs that were often present in a small concentration. After the mop-up, the polymerase screening was performed at 37 °C. The following DNA duplex template was used in experiments featuring ^{33}P-5'-labelled primer strand (denoted by an asterisk):

5' - *TGCAGGCATGCAAGCTTGGCGTA - 3'   23nt
3' - ACGTCCGTACGTTCGAACCGCATAAAAAAAAAAAA – 5'   35nt

Polymerization was performed in 20 µl of the reaction mixture at 37 °C. After certain time periods (5, 15 and 60 minutes), an aliquot of the reaction mixture was supplemented with dTTP (up to 50 µM final concentration) and the reaction was allowed to proceed for additional 5 min at the same temperature. The reactions were stopped by adding an equal volume of a STOP solution (Fermentas catalogue, item #K1711: CycleReader™ DNA sequencing kit) containing EDTA, then the products were resolved on 15 % 29:1 denaturing (7 M urea) polyacrylamide (PAA) gel run at 50 °C, dried on a

2.2.2 Incorporation of 3'-O-CEM-dTTP

Some polymerases representing different classes were found to possess markedly different efficiencies regarding the incorporation of 3'-modified nucleotides into the DNA template.

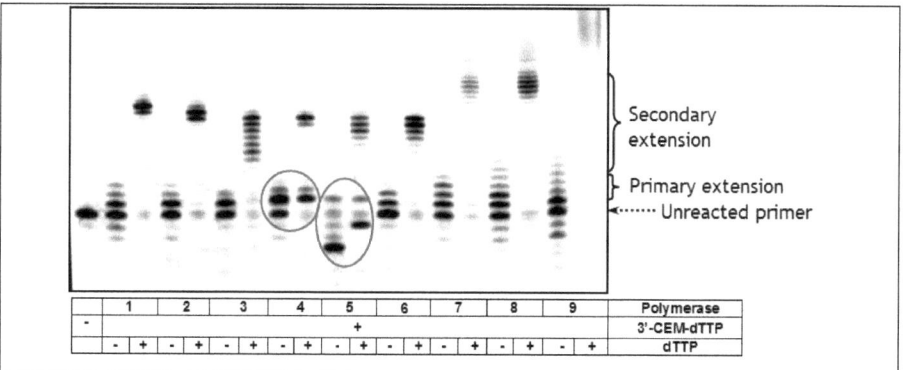

Figure 12: Primer extension by 9 polymerases (numbered from 1 to 9) using 3'-O-CEM-dTTP after 60 min incubation time

The gel picture displayed in *figure 12* was obtained from the incorporation experiment employing 3'-O-CEM-dTTP as reversible terminator. After 60 min incubation time, the incorporation of the CEM-blocked nucleotide by nine different polymerases is controlled. The addition of the naturally occurring nucleotide dTTP after the incorporation of 3'-O-CEM-dTTP results either in elimination of the modified nucleotides from the extended primer or in further primer extension by incorporation of dTTP. Here, only two polymerases (numbered 4 and 5) exhibit the capability for both primer extension and termination as indicated by the red circles. However, polymerase 5 is recognized by rather intensive exonucleolytic activity (primer shortening) and thus is barely suitable for our SBS-approach. In case of polymerase 4 the provided 3'-blocked nucleotide is not only incorporated into DNA but also terminates further extension by addition of dTTP. Regarding polymerase 4, also a small amount of (n+2)-product is observed (see weak band above (n+1)-band). To explain this, several mechanisms enabling the polymerase to avoid 3'- blocking of the primer and succeed with the DNA extension are known. One of them is the polymerase-guided direct deblocking of this group, known as "editing" activity[46]. This would result in removal of the 3'-blocking group without affecting the whole nucleotide. Another way is the removal of the whole 3'-modified terminating nucleotide through triphosphate-mediated pyrophosphorolysis[55a]. Both mechanisms are compatible with further primer extension when dTTP is present after the 3'-CEM-dTTP incorporation step.

2.2.3 Incorporation of 3'-O-CE-dTTP

Besides the 3'-modified nucleotide 3'-O-CEM-dTTP, the similar compound 3'-O-CE-dTTP was also found to be incorporated into DNA by several polymerases and terminating further primer extension. The gel picture shown in *figure 13* illustrates the incorporation of 3'-O-CE-dTTP by selected polymerases. Here, polymerases 2, 4 and 5 possess the ability to incorporate 3'-O-CE-dTTP (indicated by red circles) while the efficiencies of incorporation and termination are divergent. Polymerase 8 is also capable for some extension of the primer; however, the termination is barely detectable. Thus, polymerases 4 and 5 should be considered as the most effective in this experiment.

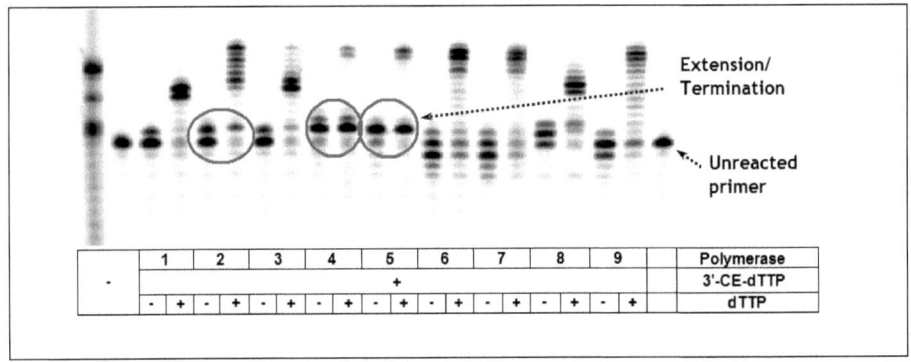

Figure 13: Primer extension by nine polymerases (numbered 1 to 9) using 3'-O-CE-dTTP after 60 min incubation time

The ability of the polymerases to incorporate 3'-O-CEM-dTTP and 3'-O-CE-dTTP appears to be dependent on the size of the 3'-blocking group: The first one bears a bigger 3'-tag and is incorporated by only two polymerases compared to six different mutants in case of the 3'-CE-labeled nucleotide. In addition, 3'-O-CE-dTTP is incorporated nearly two times fasted under identical reaction conditions than 3'-O-CEM-dTTP (see *figure 14*).

Based on the results described above, the 2-cyanoethyl (CE) group was recognized as the most promising for further work on optimization of primer extension and addressing the specificity issue for the polymerization reaction.

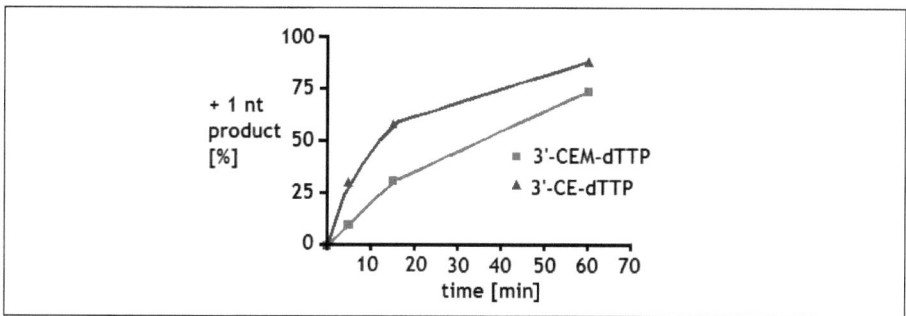

Figure 14: Primer extension efficiencies using 3'-O-CEM-dTTP or 3'-O-CE-dTTP

Fine-tuning of the reaction mixture increased the reaction speed up to the ten fold, driving the reaction of the primer extension to near completion in 5 min (not shown). This timescale is considered to be compatible with our SBS approach.

2.2.4 Concluding remarks

Several principal sources of enzymes were considered for the screening, including commercially available and Fermentas proprietary polymerases, representing all families of DNA polymerases. After initial screening tests done at Fermentas' labs using number of 3'-modified nucleotides, 23 polymerases showed acceptance towards 3'-modified dTTPs with lower or absent nucleolytic activity, both of mesophylic and thermophylic origin. A full-range screening program was executed focusing on four 3'-modified dTTP's, featuring methyldi-thioterbutyl (DTM), methylthiomethyl (MTM), (2-cyanoethoxy)methyl (CEM) and 2-cyanoethyl (CE) as 3'-protective groups albeit many more were investigated using the most promising polymerases for final evaluation, too.
As a result of these first screenings, 3'-O-CEM-dTTP and 3'-O-CE-dTTP were found to be incorporated into the DNA template by several polymerases and terminating further primer extension thus fulfilling the requirements raised. In summary, two polymerases capable to incorporate 3'-O-CEM-dTTP and six polymerases incorporating 3'-O-CE-dTTP were discovered making the 2-cyanoethyl blocking group more attractive for further enzymatic tests regarding the complete reversible terminators.

3 Goal of this PhD thesis

Based on the encouraging results from the polymerase acceptance tests with unlabeled 3'-O-modified dTTPs, the main objective of the ArraySBS project was the design and the synthesis of four complete reversible terminators. This PhD thesis covers the synthesis of the four 3'-modified key compounds needed as building blocks for the synthesis of the complete dye-labeled reversible terminators as shown in *figure 15*. Using these reversible terminators further enzymatic tests and the development of a polymerase affording the SBS proof-of-principle should be performed. The design and synthesis of the linker, the triphosphate synthesis and the dye-linker attachment will be published soon in another PhD work[55b].

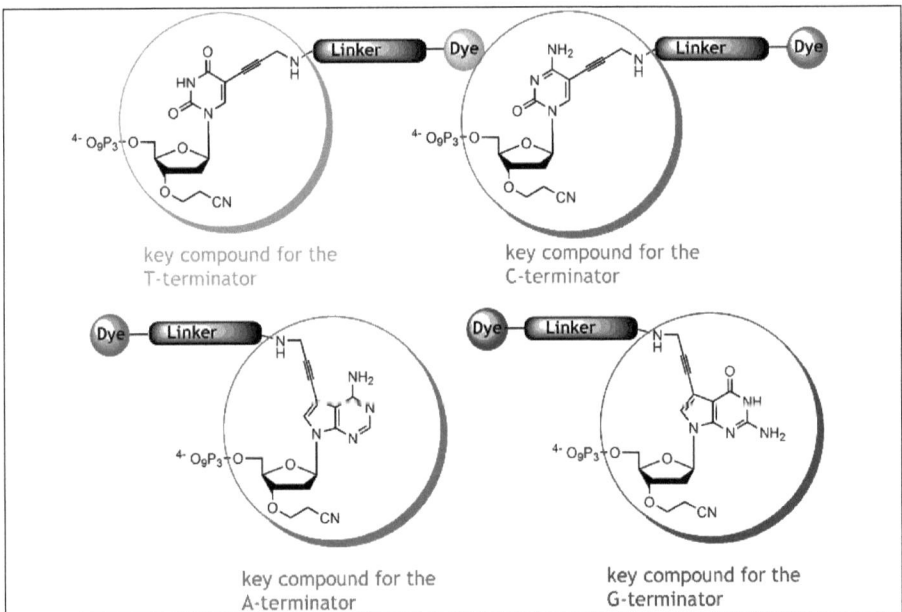

Figure 15: *The four key compounds of the complete reversible terminators*

For successful preparation of the four key compounds (highlighted by circles in *figure 15*), the following points had to be addressed within this work: For each key compound, a multi-step procedure consisting of nucleoside preparation and nucleoside modification had to be designed and put into practice. Another important issue was the spectroscopic characterization of each synthetic intermediate, which often had no spectroscopical reference data.

Another main objective of this PhD thesis was the evaluation of quantitative cleavage of the 3'-modifications like the (2-cyanoethoxy)methyl (CEM) and the 2-cyanoethyl (CE) group. Therefore a suitable experimental procedure as well as simple model compounds and appropriate cleavage reagents had to be identified. The results from the cleavage tests on a simple model compound should be applicable to the complete SBS system then, consisting of a defined template like e. g. the *p53*-suppressor gene immobilized on a chip, the four complete dye-labeled reversible terminators bearing the 3'-blocking group and a highly tolerable polymerase designed for these special nucleotides.

4 Synthesis of reversible terminators for SBS

4.1 Retrosynthesis of the four 3'-O-modified key compounds

4.1.1 Retrosynthesis of the key compound for the A-terminator

The retrosynthesis of the key compound that is needed for the preparation of the A-terminator is shown in *scheme 1*.

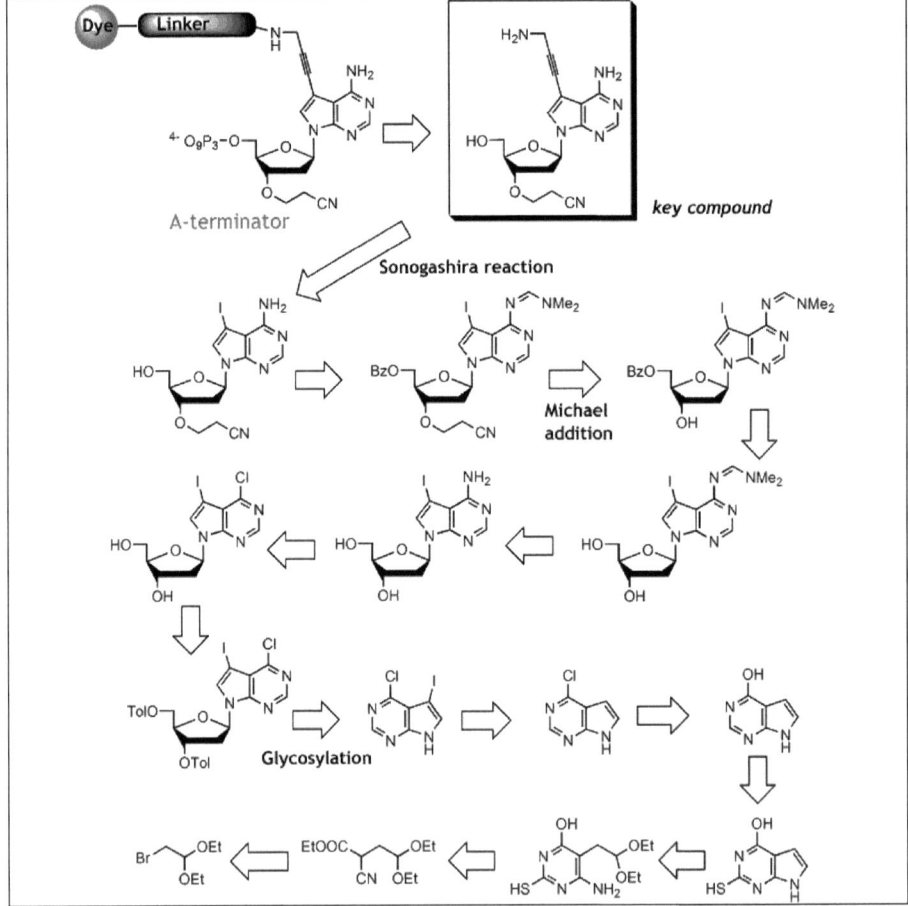

Scheme 1: Retrosynthesis of the key compound for the A-terminator

As it is depicted in *scheme 1*, the nucleobase 4-amino-5-iodo-7-(2-deoxy-β-D-*erythro*-pentofuranosyl)-*7H*-pyrrolo[2,3-*d*]pyrimidine had to be synthesized first. This means that a multi-step procedure containing the heterocyclic chemistry and a glycosylation step had to be used for building this particular nucleoside since this compound is hardly commercially available and very expensive. The heterocycle preparation, the glycosylation and the protecting group strategy form a laborious synthetic procedure, starting from gram scale and ending up in milligram scale leading to the desired key compound. The synthesis here looks straight forward at first sight, but it also had to be elaborated initially because only parts of this synthetic pathway were known in literature: For example, the preparation of the heterocycle was already well established by Davoll[56] in the 1960's, the chlorination, iodination and glycosylation were optimized by others decades later. The challenging part after the synthesis of the 5-iodo-pyrrolo[2,3-*d*]pyrimidine nucleoside was to find the shortest protecting group strategy enabling selective introduction of the cyanoethyl group to the 3'-end via Michael addition. After introduction of the propargylamine moiety to the nucleobase, the key compound should be obtained.

4.1.2 Retrosynthesis of the key compound for the G-terminator

The G-terminator was prepared in a similar manner like the A-terminator as displayed in *scheme 2a*. Here, all reactive functional groups on the heterocyclic moiety had to be protected as well as the 5' hydroxyl function for selective 3'-introduction of the CE group. This multistep protection group strategy had to be evaluated first on 2'-deoxyguanosine to give the shortest synthetic pathway with optimized yields for each step. Another big difference in the synthesis of the G-terminator, compared to the synthesis of the A-terminator, is the modified attachment of the dye-linker system[55b] that did not need the introduction of the propargylamine moiety anymore, saving two extra synthetic steps. In general, the synthesis of 2-amino-5-iodo-7-(2-deoxy-β-D-*erythro*-pentofuranosyl)-*7H*-pyrrolo[2,3-*d*]pyrimidin-4-one (see *scheme 2b*) employs a few steps more but is based on the same strategy of heterocycle preparation and glycosylation method as the one used for the synthesis of the key compound for the A-terminator.

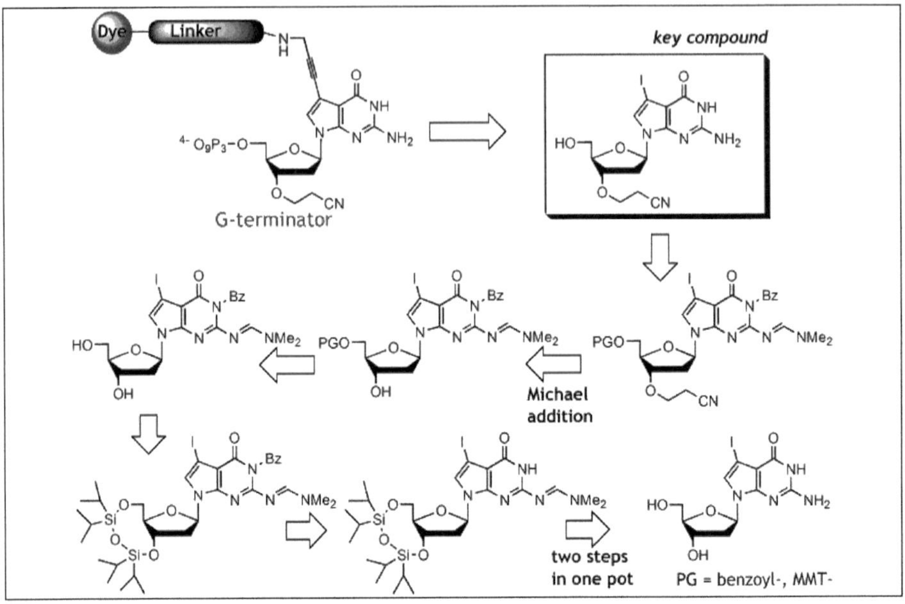

Synthesis of reversible terminators for SBS

Scheme 2a: Retrosynthesis of the key compound for the G-terminator part 1

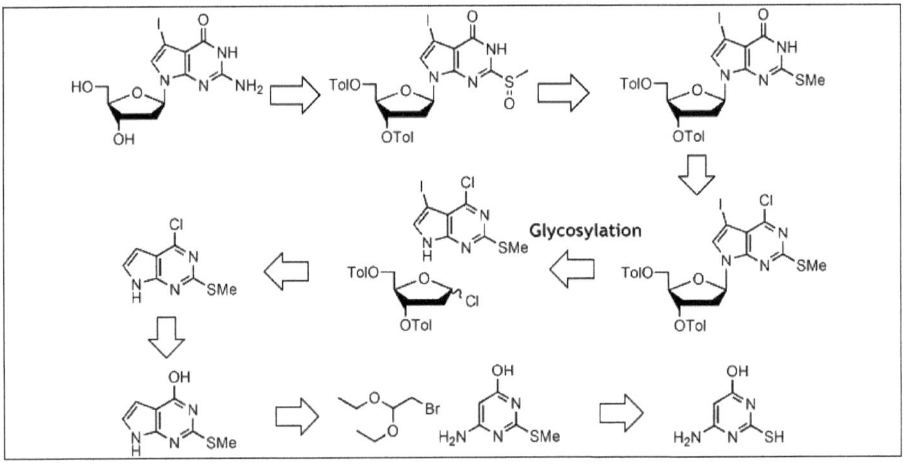

Scheme 2b: Retrosynthesis of the key compound for the G-terminator part 2

A closer look on the heterocyle preparation for the G-terminator reveals that the 2-amino group is masked as methylthio function for simplification of the glycosylation. Besides this strategy, there are several publications about the glycosylation step employing the amino-protected heterocycle. This work covers and discusses both possibilities and which one of them proved to be more successful in our hands. Aside from the nucleoside preparation, the whole protecting group strategy for enabling 3'-O-selective introduction of the CE function is elaborated within this work. In our case the preparation of the 5-iodo-pyrrolo[2,3-*d*]pyrimidine nucleoside itself was crucial due to huge loss of product during the multi-step synthesis. For this reason we decided to purchase 2 grams of this nucleoside at the final stage of the project.

4.1.3   Retrosynthesis of the key compound for the C-terminator

The retrosynthesis of the C-terminator key compound is shown in *scheme 3*.

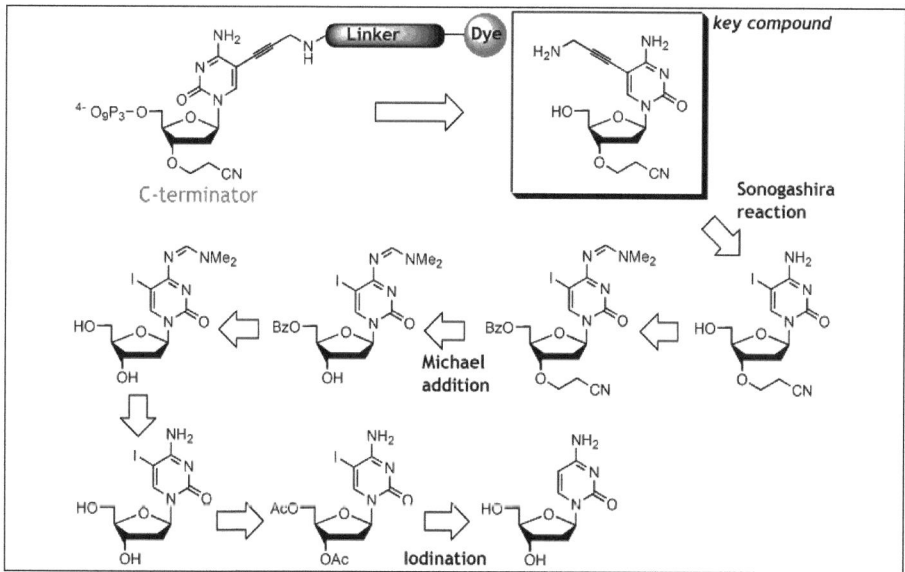

Scheme 3: Retrosynthesis of the key compound for the C-terminator

Only the iodine function has to be introduced into 5-carbon position, then a smart protection group strategy enables 3'-O-selective introduction of the CE function followed by the attachment of the propargylamine moiety. Compared to the syntheses of the A- and the G-terminator few steps less are needed to obtain the 3'-O-CE-blocked key compound with the propargylamine linker attached to the base moiety. The preparation of the C-terminator is shorter and more economical than the syntheses of the pyrrrolo[2,3-d]pyrimidine base-terminators delivering sufficient amount for further synthetic steps[55b].

4.1.4 Retrosynthesis of the key compound for the T-terminator

The synthesis of the T-terminator is the shortest and simplest one compared to the methods for the preparation of the other three terminators as shown in the retrosynthetic analysis (*scheme 4*).

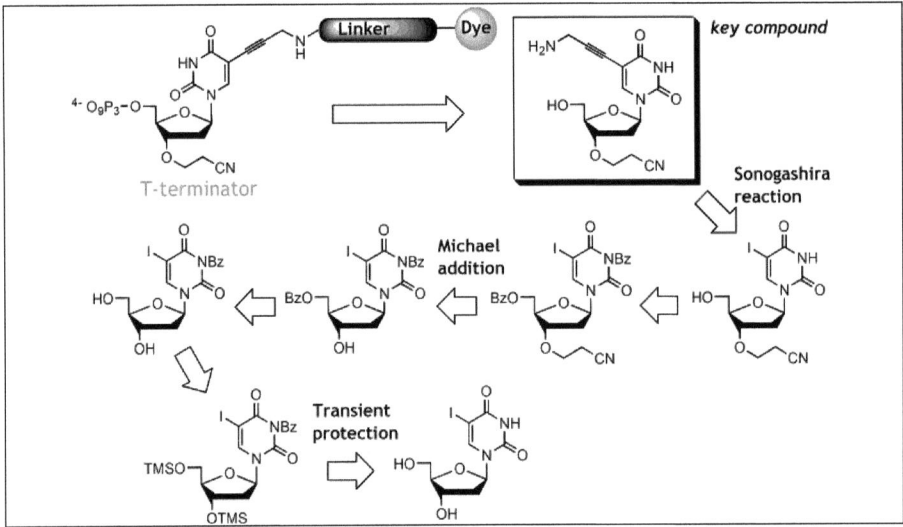

Scheme 4: Retrosynthesis of the T-terminator

One big advantage is the commercial availability of 5-iodo-2'-deoxyuridine providing us with enough starting material for the preparation of the whole terminator in larger amounts. Here an elegant protecting group strategy was evaluated for enabling selective introduction of the 2-cyanoethyl function on the 3' hydroxyl (see *scheme 4*). As described before, the CE function can only be introduced selectively if any other acidic protons are protected during the Michael reaction. The Sonogashira reaction with subsequent liberation of the amino function enables the dye-linker attachment[55b]. These four retro-syntheses presented in this work are four possible synthetic solutions for the preparation the desired key compounds. We have chosen these four strategies because they meet the criteria of the shortest syntheses with yield optimization for many of the synthetic steps.

4.2 Strategy for selective 3'-alkylation of 2'-deoxyguanosine

4.2.1 Dialkylation of partially protected 2'-deoxyguanosine

As already presented in the retrosynthetic analysis of the G-terminator, a smart protecting group strategy for the selective introduction of the 2-cyanoethyl function on the 3'-position had to be developed. As test nucleoside for this evaluation, we used 2'-deoxyguanosine instead of the expensive 2-amino-5-iodo-7-[2-deoxy-β-D-*erythro*-pentofuranosyl]-*7H*-pyrrolo[2,3-*d*]pyrimidin-4-one.

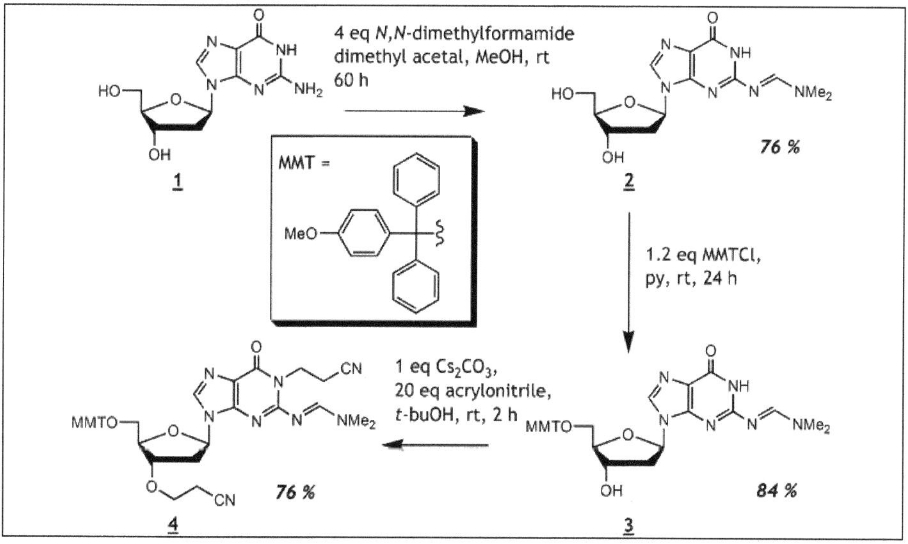

Scheme 5: Dialkylation of partially protected 2'-deoxyguanosine

One obstacle in selective alkylation via the Michael addition[57a] of the 3'-position is the high reactivity and therefore poor selectivity of acrylonitrile that is added to any nucleophilic function of 2'-deoxyguanosine. We could demonstrate the low selectivity by carrying out the following test alkylation, see *scheme 5*. The dialkylated nucleoside **4** was formed as single product as confirmed by mass spectrometry and ^{1}H-NMR analysis (mass spectrum see Annex). The reason for this may be explained by having a closer look at the reaction on the nucleoside as shown in *scheme 6*.

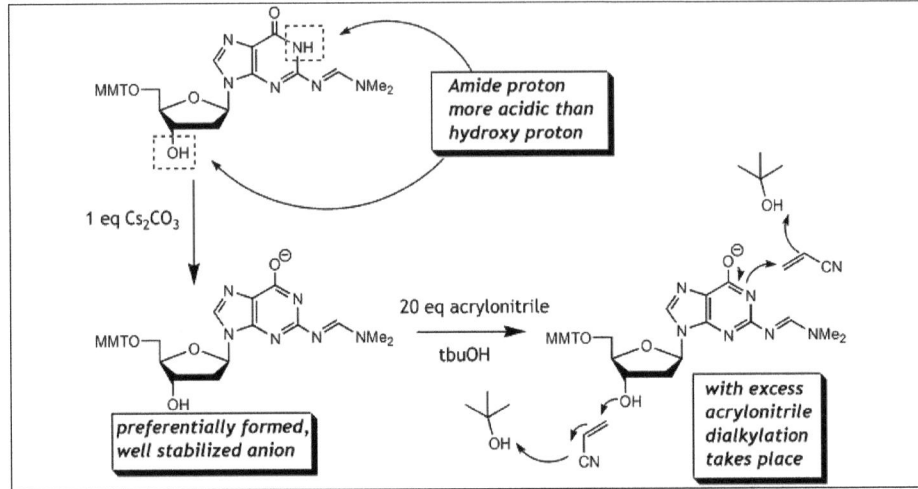

Scheme 6: Concurring sites during Michael addition

Cesium carbonate acts as base and preferentially deprotonates the more acidic amide proton. The resulting anion is well stabilized and is alkylated with acrylonitrile. The attempt to remove the CE group selectively from the *N3*-position of compound **4** was not successful: this moiety could not be cleaved without affecting the 3'-O-(2-cyanoethyl) group by treating the dialkylated nucleoside with TBAF/THF. As a conclusion from this experiment we decided to change our synthetic strategy for enabling selective 3'-alkylation.

4.2.2 Selective 3'-alkylation of fully protected 2'-deoxyguanosine

In order to avoid the dialkylation any nucleophilic function on 2'-deoxyguanosine was protected except the 3'-hydroxy group. The formamidino moiety on the exocyclic amino function showed to be stable enough during further synthetic steps, also the MMT-function on the 5'-position could be introduced in good yield, but still the 3-amido function remained unprotected and could not be selectively protected in the presence of the free 3'-hydroxy function. Transient protection of the 3'-hydroxy function with chlorotrimethylsilane (TMSCl) and subsequent benzoylation on 3-nitrogen position failed, so we had to use a different approach, as it is demonstrated in *scheme 7a*.

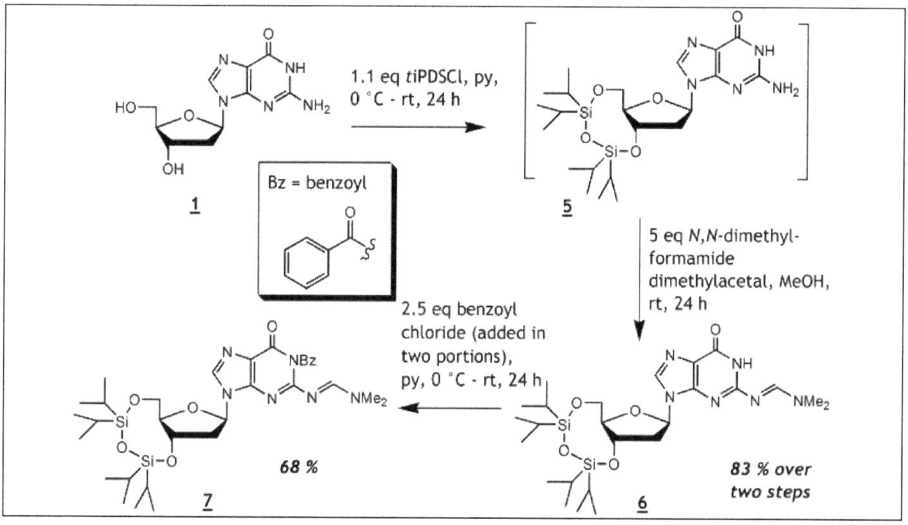

Scheme 7a: Transient protection for selective benzoylation of the N3-position

For quantitative protection of the hydroxyl functions, we discovered that the Markiewicz group[58] was the appropriate choice: It is not as labile as the TMS groups, but on the other hand not too stable like e. g. *tert*-butyltrimethylsilyl groups. Another big advantage of the Markiewicz protection is its quantitative and highly selective introduction, even though the isolation of the Markiewicz-protected 2'-deoxyguanosine is difficult due to the fact that the nucleoside tends to form emulsions during the aqueous workup. In order to achieve a high yield of protected nucleoside **6** ready for selective benzoylation, the formamidino function was introduced *in situ*: After quenching the first protection reaction with methanol and evaporation of the solvent, nucleoside **5** was not isolated but immediately treated with N,N-dimethylformamide dimethyl acetal in methanol giving the desired compound **6** in an excellent yield of 83 % over two steps. The benzoylation of the 3-nitrogen function gave in moderate yield the fully protected compound **7**. We also suggested other protecting groups for the 3-nitrogen position, even though there were only few alternatives published[59]. One example we tried before the benzoyl protection was the triisopropylbenzenesulfonyl group readily introduced to the 4-carbonyl position[59c].

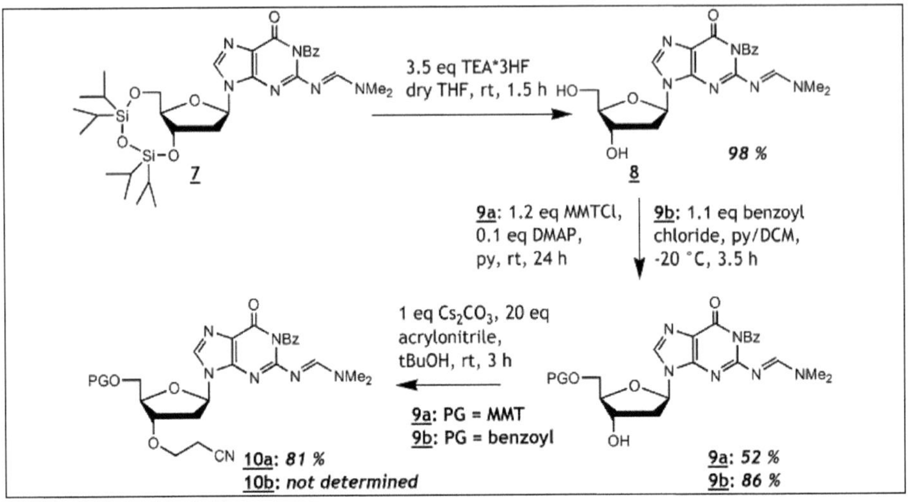

Scheme 7b: Preparation of fully protected 3'-O-CE-2'-deoxyguanosine

Nevertheless this group was too labile and substituted easily with good nucleophilic bases like ammonia[59c] or fluoride, making it useless for our synthetic strategy. Also the benzyl protection[60] of the 4-oxygen position is not the best choice here: it is only Pd-catalyzed and removable under harsh conditions[60] and therefore too stable for our purposes. In contrast to that, the benzoyl group is quantitatively cleavable in aqueous ammonia, but stable during the next synthetic steps (see *scheme 7b*). The Markiewicz protecting group was removed quantitatively under mild conditions, *i. e.* deprotection with triethylamine trihydrofluoride in THF[57a] instead of TBAF in THF. The resulting nucleoside **8** had to be dried very well for the next step, the selective 5'-hydroxy protection. Here we found a big difference in reaction efficiency depending on the type of protecting group: Contrary to our expectation, the MMT-protection step[57b,c] (see *scheme 7b*) was quite ineffective so that half of the starting material **9a** was isolated even though DMAP was used as catalyst for this reaction. In contrast to that, the 5'-benzoylation at -20 °C as an alternative protection delivered compound **9b** in a satisfying yield of 86 %. The Michael addition for introduction of the CE group gave the opposite results: Here, the reaction with MMT-protected nucleoside **9a** was more efficient yielding 81 % product **10a** than the Michael addition performed with the 5'-benzoyl protected starting material **9b**. Besides that, the obtained compound **10b** was hard to purify and therefore used without further purification for the deprotection step. The removal of both benzoyl- and MMT-protection groups gave an unexpected outcome as it is illustrated in *scheme 7c*.

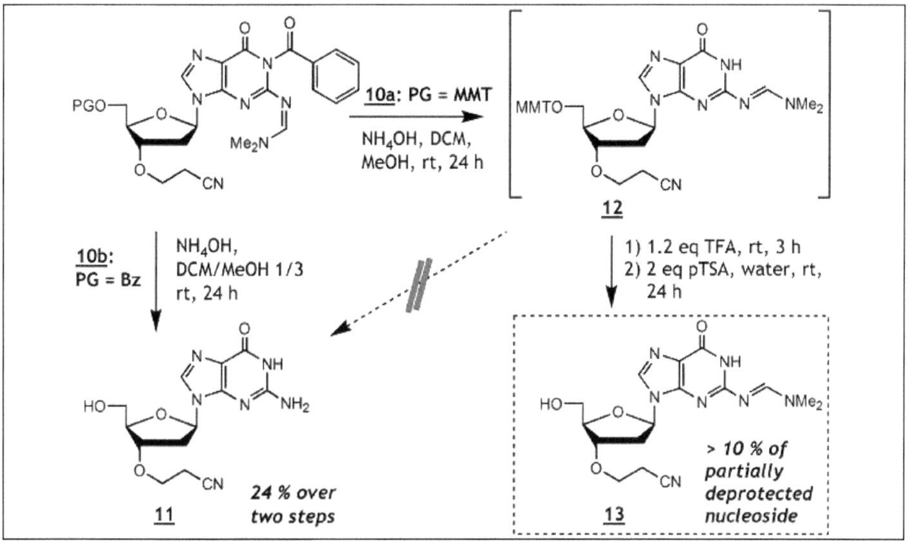

Scheme 7c: Deprotection of benzoyl- and MMT-protected nucleosides

At first, the attempt to remove the *N3*-benzoyl and the formamidino group from the MMT-protected nucleoside **10a** was ineffective: The formamidino group was hardly removable and extremely stable while the benzoyl function was cleaved in high amount. The solvents and the ammonia were then removed under reduced pressure and the crude product **12** was characterized via mass spectroscopy. In order to prevent loss of material, it was decided to skip the purification of the product and subject it directly to the next deprotection step for MMT-removal. Employing an excess of pTSA did not lead to the desired deprotection first, but the addition of a slight excess of TFA started the reaction. Unfortunately the reaction did not proceed as expected: The resulting products detected via TLC showed a high polarity complicating the isolation and purification of the crude product. The main fraction isolated after a short flash chromatography was analyzed with ESI mass spectrometry. The spectrum is shown below in *figure 16* revealing the peak of the deprotected nucleoside **13**, but also signals from lower masses.

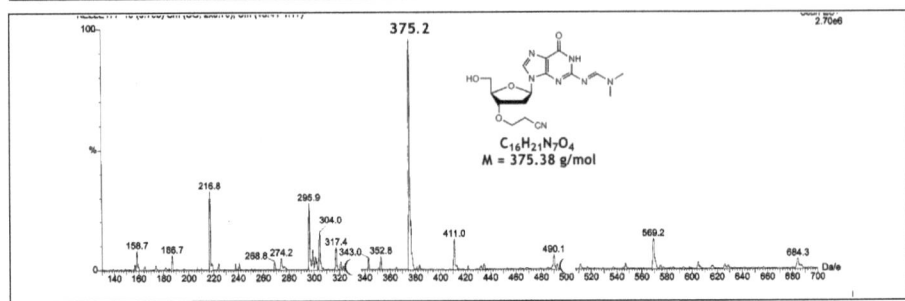

Figure 16: ESI(+)-mass spectrum of the product from the MMT-deprotection

We assume that nucleoside **12** hydrolyzed under these strong acidic conditions. Unlike nucleoside **10a**, the 5'-benzoyl protected compound **10b** was deprotected with aqueous ammonia delivering the desired product **11** in moderate yield. The design of the described test synthesis of 3'-O-(2-cyanoethyl)-2'-deoxyguanosine helped us at last to achieve the ultimate goal, the efficient preparation of 2-amino-7-[2-deoxy-β-D-*erythro*-pentofuranosyl]-7H-pyrrolo[2,3-d]pyrimidin-4-one.

4.3 Synthesis of the 3'-modified key compounds

4.3.1 Synthesis of the pyrrolo[2,3-*d*]pyrimidine moiety: Strategy 1

For the preparation of both the A- and the G- terminator, the natural nucleosides 2'-deoxyadenosine and 2'-deoxyguanosine are not applicable due to the fact that it is impossible to attach a linker-dye system at their 5-position. For enabling such an attachment, both nucleobases have to be iodinated pyrrolo[2,3-*d*]pyrimidine nucleosides. Herein and also in the following chapters we have chosen the systematic numbering for all compounds formed during the syntheses of the A- and the G-terminator (see *figure 17*).

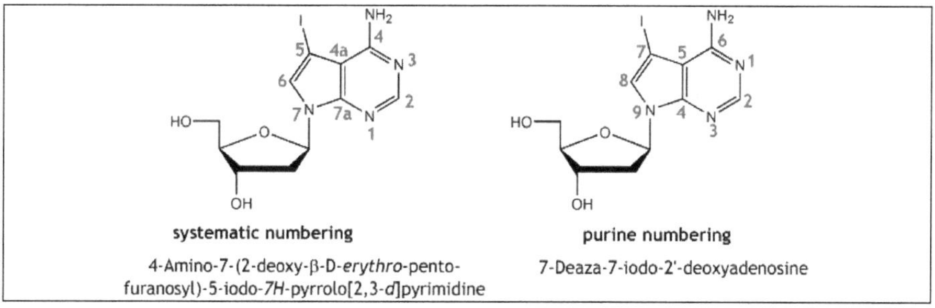

systematic numbering

4-Amino-7-(2-deoxy-β-D-*erythro*-pento-furanosyl)-5-iodo-7H-pyrrolo[2,3-*d*]pyrimidine

purine numbering

7-Deaza-7-iodo-2'-deoxyadenosine

Figure 17: Systematic numbering and purine numbering

The most common synthetic strategy reported in literature is to prepare the 5-iodo-pyrrolo[2,3-*d*]pyridine derivative following Davoll's protocol[56] first and glycosylate the heterocycle afterwards. One attempt for the synthesis of the nucleobase moiety we have chosen is shown in *scheme 8*. We selected 2-amino-6-hydroxy-2-mercapto-pyrimidine **14** as commercially available starting material here due to the fact that 2-amino-6-hydroxypyrimidine is hardly available and of high cost.

Therefore we started the synthesis with the methylation[61] of compound **14**: Iodomethane acts as a very selective sulfur methylation agent under alkaline conditions in water. The thiol-function is deprotonated with sodium hydroxide and undergoes a nucleophilic substitution to form the methylated heterocycle.

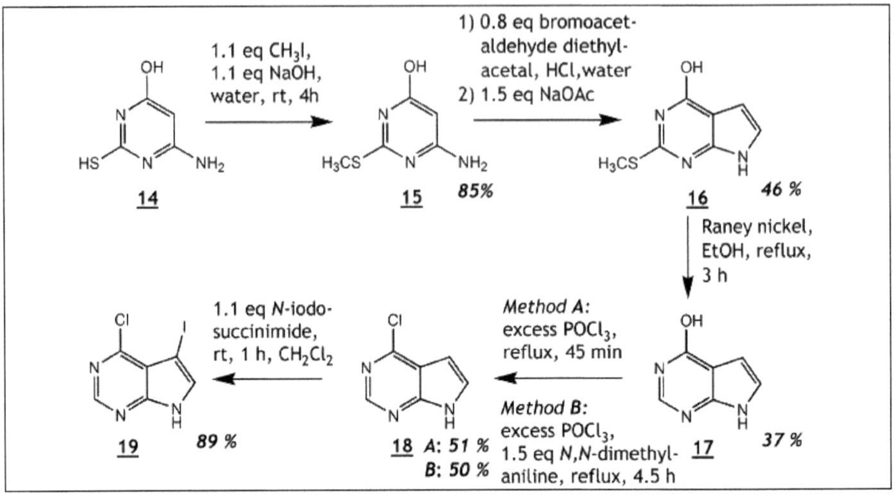

Scheme 8: Preparation of the pyrrolo[2,3-d]pyrimidine moiety, method A

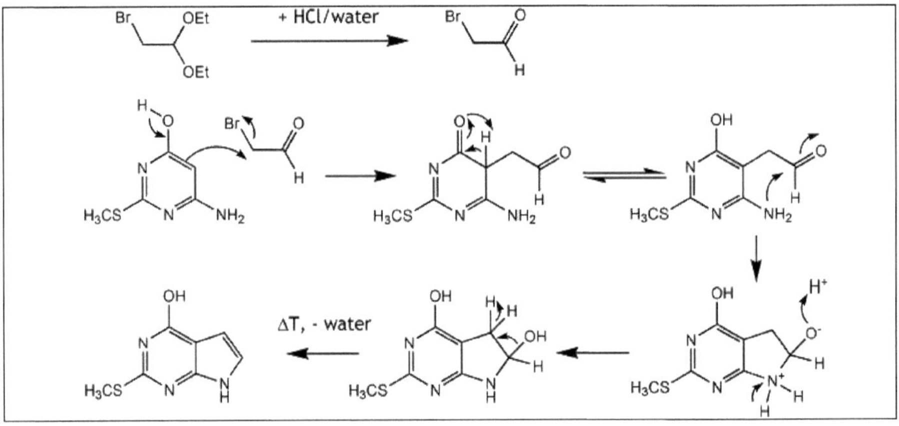

Scheme 9: Mechanism of the condensation reaction

Converting product **15** into compound **16** was achieved by using Barnett's method[62], but the resulting yield was only moderate around 50 % due to workup problems: During the alkylation of the pyridine ring with bromoacetaldehyde diethylacetal in water the product precipitated not only as crystals, but as well as a yellow sticky mass. This kind of polymerized product was often formed in large quantities if the suspension was too concentrated and not stirred vigorously. The reason may be that the open heterocycle, which exists during the alkylation with bromoacetaldehyde, does not only react intramolecular, but intermolecular as well. It was not possible to prevent this problem completely even while running the reaction in high dilution. The mechanism of the reaction is shown in *scheme 9* illustrating the condensation between the pyrimidine ring and the bromoacetal: In an extra flask, the acetal-protected bromoaldehyde was activated with hydrochloric acid. The acidic solution was then buffered with sodium acetate and added to the suspension of compound **15**. With heating and strong stirring, the condensation reaction could take place and the heterocycle **16** was formed. The 2-methylthio function of heterocycle **16** was selectively removed then by using Raney nickel under conditions as described in literature[66]. It is recommended that the reaction mixture should not reflux longer than 3 h, otherwise the heterocycle decomposes. The nickel aluminum alloy was removed via filtration over Celite while the solvent was still hot. This technique prevented that most of the product **17** got stuck in the filter cake and could be isolated in moderate yield after concentration of the filtrate.

For further synthesis, it was useful to mask the 4-hydroxy moiety with a chlorine function as reported in literature by Seela et al.[63]. The chlorination with $POCl_3$ had to be carried out carefully without overcoming the reaction time of roughly 4.5 h. This reaction often led to low yields: Using the very protic and reactive phosphorus oxychloride as reagent and solvent made it impossible to monitor the reaction via TLC. A reaction time of less than 45 min did not convert the starting material **17** into the desired product **18**. We also used *N,N*-dimethylaniline once for activating phosphorus oxychloride[64] and ran the reaction longer (between 2 and 4.5 h), but the yield of the desired compound **18** did not increase. According to this unhandy chlorination, the extraction of the product from the acidic aqueous layer was also not optimal. The product showed only low solubility in any organic solvent like methylene chloride, *n*-hexane, chloroform, diethyl ether etc., complicating its isolation.

The most efficient extraction solvent was ethyl acetate, but still big amounts of this solvent were needed for the extraction of product **18**. In some cases the yield was only around 30 % of pure and crystalline product and although this reaction was repeated a few times the yield did never surpass roughly 50 %. In contrast to that, the selective iodination in 5-carbon position worked quite well. In principle, both 5- and 6-carbon may abstract the electrophilic iodine, but if only 1.1 equiv. of *N*-iodosuccinimide are added, the 5-carbon position is preferentially iodinated. The reaction was carried out in dry methylene chloride or DMF and gave always compound **19** as single product with yields ranging from 80 to 90 %. Looking at synthetic strategy 1 for the preparation of heterocycle **19** an overall yield of roughly 7 % over 5 steps was achieved. This means the synthetic strategy was still not optimized and had to be started in large gram-scale for isolation of sufficient amount of product.

4.3.2 Synthesis of the pyrrolo[2,3-d]pyrimidine moiety: Strategy 2

Besides the first strategy starting with the Barnett method, we tried another more conventional synthetic route shown in *scheme 10*. The second route has one reaction step more and uses the first steps as reported in Davoll's protocol[56].

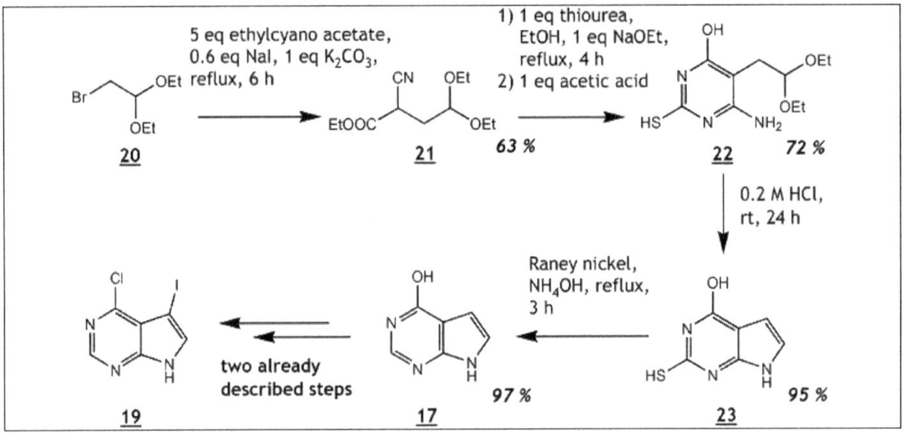

Scheme 10: Preparation of the pyrrolo[2,3-d]pyrimidine moiety, method B

The synthesis started again with bromoacetaldehyde diethylacetal that forms the α-cyanoester **21** with ethylcyano acetate under alkaline conditions and in moderate yields of about 60 %. This α-cyanoester was condensated then with thiourea to give compound **22** in moderate yield. The following ring closure always worked quantitatively in acidic medium. The 2-thiol function was also removable with Raney nickel, but this time we ran the reaction in aqueous ammonia[56] for forming a clear solution of starting material **23**. The chlorination as well as the iodination of compound **17** was carried out as already described for synthetic route 1. The overall yield of synthetic strategy 2 was 19 % over 6 steps. Comparing synthetic strategy 1 (overall yield ca. 7 %, see chapter 4.3.1.) with synthetic strategy 2, the desired heterocycle **19** was delivered in more than double amount. Although synthetic strategy 2 employs one step more than strategy 1, it reveals a much higher efficiency during the desulphurization and therefore is the more efficient one.

4.3.3 Synthesis of 4-amino-7-[2-deoxy-β-D-*erythro*-pentofuranosyl]-5-iodo-7*H*-pyrrolo[2,3-*d*]pyrimidine

With the halogenated pyrrolo[2,3-*d*]pyrimidine moieties **18** and **19** in hand, one part of the nucleoside was already successfully prepared. The second part needed for building the adenosine derivative was the sugar moiety.

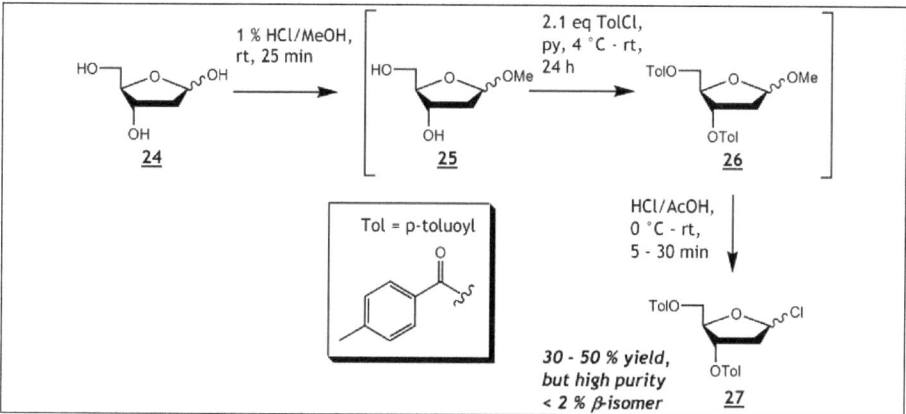

Scheme 11: Preparation of the sugar moiety according to Rollard et al.[67]

The sugar compound 2-deoxy-3,5-di-*O*-p-toluoyl-α-D-*erythro*-pentofuranosylchloride **27** (see *scheme 11*) was synthesized by using the conventional method of Rolland *et al.*[67] This three-step procedure started with a simple acetal formation converting 2-deoxy-D-ribose **24** in 1 % methanolic hydrogen chloride at room temperature into compound **25**. After quenching and aqueous workup, the residual oily acetal **25** was coevaporated three times with pyridine, dissolved in dry pyridine and cooled down to 0 °C. Roughly 2.1 equiv of *p*-toluoyl chloride were added drop-wise within 1 h and after stirring the mixture overnight at 4 °C the reaction was complete. After an aqueous workup and evaporation of the solvent, the resulting syrup containing fully-protected sugar **26** was diluted in 40 ml acetic acid. In an Erlenmeyer flask, the HCl generating mixture was prepared by addition of acetyl chloride to acetic acid on cooling. The optimal temperature range for this was around 5 - 15 °C, then a small amount of water was added to that mixture starting the generation of HCl. The diluted sugar was subsequently added to this mixture via pipette on cooling. The colorless crystals consisting mainly of the α-isomer of **27** precipitated immediately after completion of sugar addition. They were subsequently filtered off, washed with dry diethyl ether for acid removal and dried in vacuum. This kind of fast handling made it possible to isolate the desired α-isomer in high purity: Usually only less than 1 or 2 % consisted of the β-isomer, although the overall yield of the desired compound **27** is often low. The high isomeric purity was necessary due to the difficult purification of the chloro sugar. This compound is very sensitive to humidity so that only crystallization in tetrachlorocarbon or chloroform can be employed as purification step without separating both isomers.

In our case, the chloro sugar was used without further purification for the glycosylation reaction with the base moiety like compound **18** or **19**. In principle, one can chose different glycosylation methods: One possibility is the method employing sodium hydride[64,68f] (60 % in mineral oil) as strong base in dry acetonitrile, like it is shown in *scheme 12a* (method C).

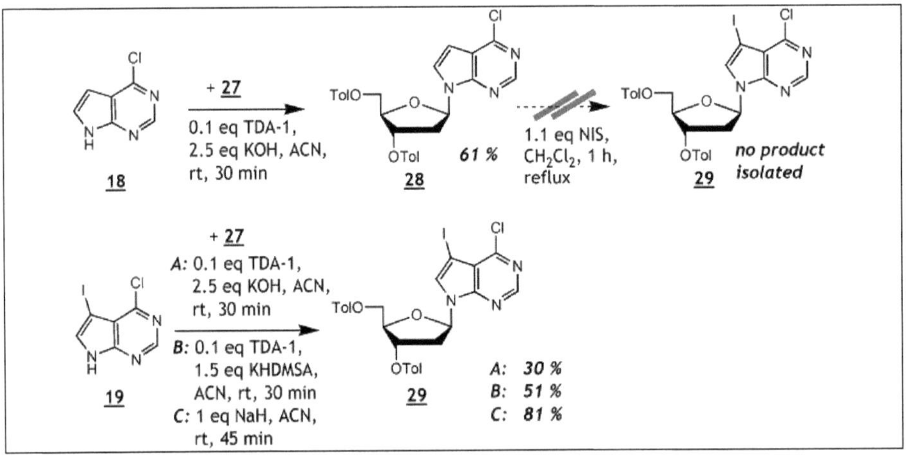

Scheme 12a: The glycosylation method according to Seela[68]

An alternative to this is the phase-transfer glycosylation which Seela et al. used several times for the preparation of different nucleobases[63,65,68a-e] (see *scheme 12a* methods A and B). The phase-transfer catalyst Seela group often used was tris(3,6-dioxaheptyl)amine (TDA-1): TDA-1 complexes the potassium cation from potassium hydroxide and releases the free alkaline anion to deprotonate the heterocycle **18** (or **19**) in acetonitrile. Subsequently the chloro sugar **27** is added to the mixture, and the anionic heterocycle reacts immediately at its 7-nitrogen with it in a nucleophilic substitution-type reaction.

The chloride function of the sugar can readily be replaced with inversion of the configuration. That means pure α-isomer-containing chloro sugar **27** is converted into pure β-isomer of nucleoside **29**. Characterization of these isomers was achieved by an ROESY-NMR experiment. If the sugar consisted of both isomers, a mixture of α- and β-nucleosides was obtained after glycosylation. This issue makes it so important that nearly α-isomeric pure sugar compound **27** is applied. We achieved the highest yields of nucleoside **29** with glycosylation method C[64,68f] (see *scheme 12a*): In this case, yields between 75 and 90 % of the desired β-isomer were obtained. Compared to this, the phase-transfer catalyzed glycosylation had one big disadvantage when using potassium hydroxide as base: After the hydroxyl anion had deprotected the heterocycle, it turned into water which led to sugar hydrolysis. In order to prevent this, potassium hydroxide was substituted by using potassium hexamethyldisilylamide (KHDMSA). The yield of the desired α-isomer increased from 30 % to 51 %, but still was not satisfying. One reason for this might be the lowered reactivity of the heterocycle anions: Regarding the nucleophilicity, heterocycle **19** is even less nucleophilic than heterocycle **18** caused by the electron-withdrawing 5-iodine substituent.

Contrary to our assumption, the introduction of iodine in 5-carbon position of nucleoside **28** after glycosylation was not possible in our hands. For prevention of this problem, it was decided to introduce the iodine into the pyrrolo[2,3-*d*]pyrimidine moiety first. At that stage, only two more steps in the synthesis of 4-amino-7-[2-deoxy-β-D-*erythro*-pentofuranosyl]-5-iodo-*7H*-pyrrolo[2,3-*d*]pyrimidine **31** were needed. One of them was the cleavage of the *p*-toluoyl protecting groups (see *scheme 13*). This was achieved in nearly quantitative yield by simply stirring nucleoside **29** in aqueous methanolic ammonia at room temperature overnight. The chlorine in 4-carbon position is very sensitive to nucleophiles, so we had to be aware of the reaction temperature not surpassing 30 °C. At elevated temperatures, the chlorine is partially replaced by a methoxy group resulting in formation of the undesired nucleoside **32**.

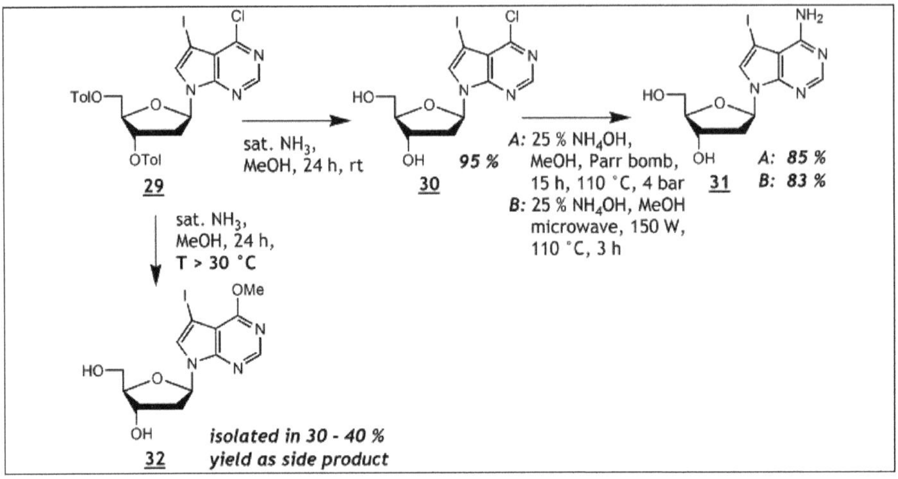

Scheme 13: Deprotection and conversion of the 4-chloro function

For the aminolysis of the chlorine function, the conventional autoclave method was used[68a]. For that purpose the starting material **30** was dissolved in methanol and aqueous ammonia, sealed in a Parr autoclave and heated several hours under the pressure being generated while heating. This reaction was also tried twice in microwave which led to shorter reaction times. Under conditions B depicted in *scheme 13* it was possible to reduce the reaction time to 3 h. Although the aminolysis worked quite well employing both methods, no further microwave-assisted trials were done due to handling problems by using methanolic ammonia. This aggressive solvent tends to spill over in the microwave apparatus so that parts of it may get affected and broken. For bigger amounts of starting material, the Parr bomb method was the more applicable one. Using the most efficient steps of the routes shown in *schemes 12a* and *13*, we achieved an optimal overall yield of 67 % of 4-amino-7-[2-deoxy-β-D-*erythro*-pentofuranosyl]-5-iodo-7*H*-pyrrolo[2,3-*d*]pyrimidine **31** over three steps, starting from the glycosylation until deprotection and aminolysis.

4.3.4 Synthesis of 4-amino-5-[3-amino-prop-1-ynyl]-7-[3-O-(2-cyanoethyl)-2-deoxy-β-D-*erythro*-pentofuranosyl]-7*H*-pyrrolo[2,3-*d*]pyrimidine

In order to enable the attachment of a dye-linker system on nucleoside **31**, a propargylamine moiety had to be introduced to the 5-carbon position. In addition to that the 3'-end of the nucleoside had to be blocked with the 2-cyanoethyl function (see *scheme 14a*). For the 3'-O-selective insertion of the CE group it was essential that all acidic protons of the nucleoside **31** except the 3'-hydroxy function itself were protected. We employed the formamidino group for protecting the exocyclic 4-amino function[69]. The introduction of this group was highly amino-selective, furnishing compound **33** in good yield without formation of any side-products. Compared to this the 5'-O-benzoylation of the deoxyribose moiety could only reach high 5'-selectivity if the reaction was carried out at low temperatures[70]: Here, the best yield of the desired nucleoside **34** was achieved by dropwise addition of approximately 1 equiv benzoyl chloride, diluted with dry dichloromethane, at -20 °C in dry pyridine.

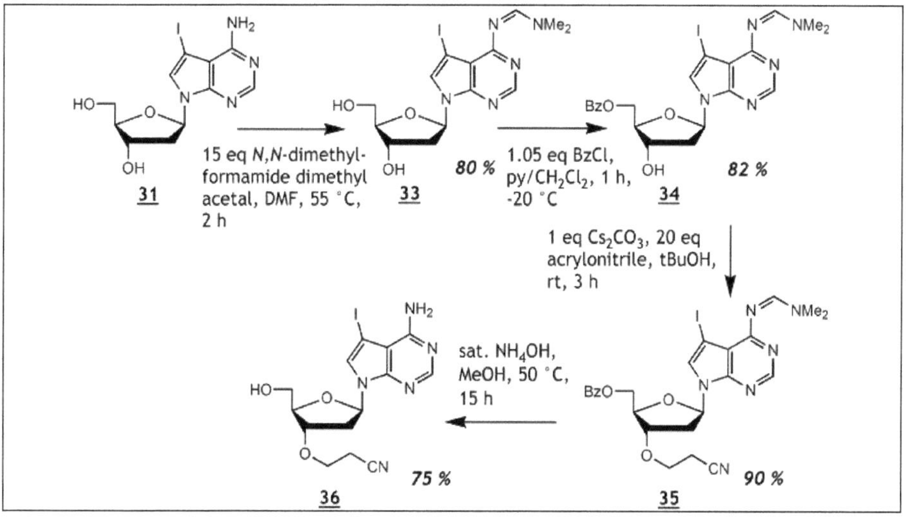

*Scheme 14a: The 3'-O-selective introduction of the CE function on nucleoside **31***

The resulting nucleoside **34** showed excellent solubility in the *tert*-butanol/acrylonitrile solvent system employed for the Michael reaction[57a]. For complete conversion of the starting material it was necessary to apply a certain experimental setup: The suspension caused by the insoluble cesium carbonate was vigorously agitated in an Erlenmeyer flask (instead of a common used round flask) with a triangle stirring bar. This geometry of the reaction vessel enabled sufficient mixing and activating of the inorganic base. Besides that, it avoided the polymerization of the excess acrylonitrile during the reaction.

After isolation of the cyanoethylated nucleoside **35** both protecting groups could be removed with a mixture of saturated methanolic ammonia and concentrated aqueous ammonia.

The resulting compound **36** underwent a Sonogashira reaction[72] for the introduction of the propargylamine function on 5-carbon position[73] (see *scheme 14b*).

In principle, this reaction was always carried out in the same manner: The nucleoside **36** was dissolved in dry DMF and treated with 5 equiv TEA, degassed three times, then allowed to warm up to room temperature and treated with 0.2 equiv copper iodide, 0.1 equiv tetrakis(triphenylphosphine)palladium and with 2 equiv of trifluoroaceto-3-aminopropargyl.

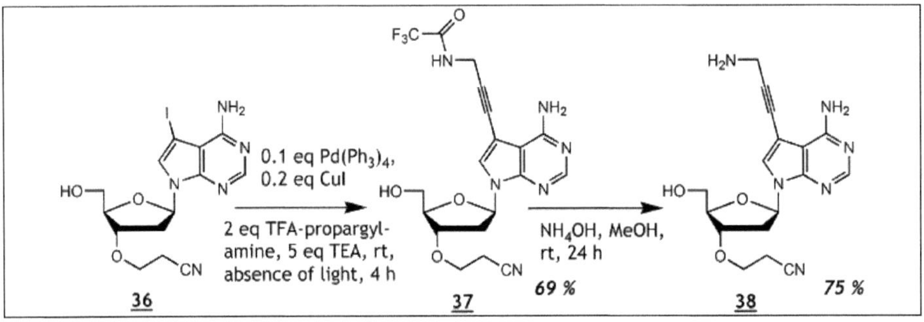

Scheme 14b: Final steps for preparation of key compound **38**

The mixture was stirred under argon with absence of light until all starting material was consumed, which was the case after 4 h reaction time. Once the mixture was stirred overnight, but prolongation of the reaction time did not increase the yield.

The mechanism[74] of the Sonogashira coupling is shown in *scheme 15*. The palladium catalyst, a Pd(0) or Pd(II) species, inserts into the carbon-halogene bond of the aromatic system (*step 1*: oxidative addition). The alkyne is deprotonated with an excess base (here triethylamine, see *scheme 15, step 2b*) and gets negatively charged resulting in binding to the copper cation from CuI (co-catalyst). The activated alkyne is transferred via transmetalation to the Pd-complex regenerating the copper salt at the same time (see *step 2a*). The Pd-complex now carries both the alkyne and the aromatic system, forming a new carbon-carbon bond after *cis/trans*-isomerization (*step 3*) and reductive elimination (*step 4*). During the last step, the product is released and the Pd-catalyst regenerated - another coupling cycle can start.

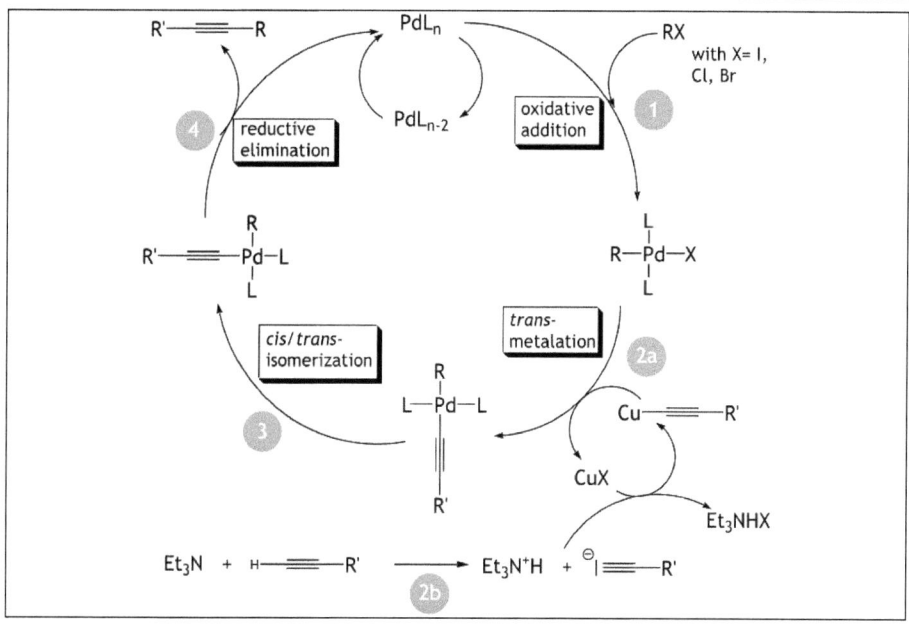

Scheme 15: Mechanism of the Sonogashira coupling[72,74]

As also described in literature[69,73], this palladium-catalyzed cross coupling reaction gave compound **37** in good yield (70 %) by employing the ratio of 2/1 copper iodide/palladium catalyst (Pd(0)-species). The TFA function is easily cleaved off with aqueous ammonia in methanol to form the target key compound **38** in moderate yield.

4.3.5 Synthesis of 2-amino-7-[2-deoxy-β-D-*erythro*-pentofuranosyl]-5-iodo-7*H*-pyrrolo[2,3-*d*]pyrimidin-4-one: Strategy 1

Several attempts for the synthesis of 2-amino-7-[2-deoxy-β-D-*erythro*-pento-furanosyl]-5-iodo-7*H*-pyrrolo[2,3-*d*]pyrimidin-4-one are published[68a,68c,75,76]. The best evaluated methods are known from Seela *et al.* who used quite similar strategies like the ones for the preparation of 4-amino-7-[2-deoxy-β-D-*erythro*-pentofuranosyl]-5-iodo-7*H*-pyrrolo-[2,3-*d*]pyrimidine **31**. One big difference of the guanosine compound is its solubility: It tends to crystallize much more than the adenosine analog and is therefore often insoluble in various organic solvents like e.g. methylene chloride. As a result of this, various approaches had been tested until a suitable synthetic route was found. Similar to nucleoside **31**, the guanosine analog cannot be prepared directly from 2'-deoxy-guanosine. Here also the 5-iodo-pyrrolo[2,3-*d*]pyrimidine moiety had to be prepared first, followed by glycosylation and 3'-modification in order to obtain the key compound for the G-terminator preparation.

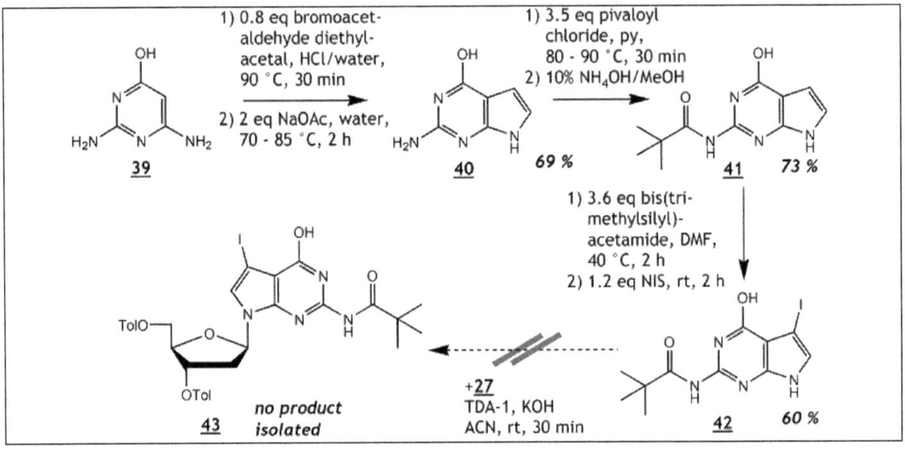

Scheme 16: First trial to synthesize the desired nucleoside **54**

The first trial to prepare the deazapurine heterocycle **42** with subsequent glycosylation is shown in *scheme 16*. Starting from commercially available 2,4-diamino-6-hydroxy-pyrimidine **39**, the pyrrolo ring could be attached in moderate yield using Barnett's method.

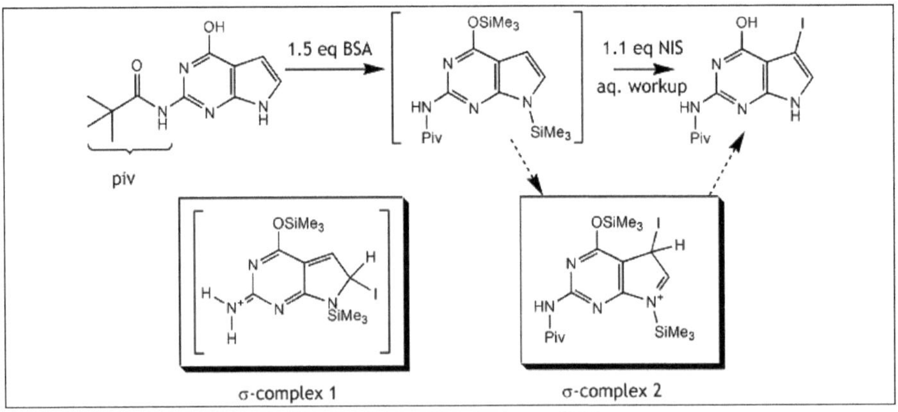

Scheme 17: Mechanism of selective iodination of 5-carbon position[62,78]

As an amino-protecting group for the subsequent iodination, the pivaloyl moiety was chosen[76] and introduced by treating heterocycle **40** with an excess of pivaloyl chloride at elevated temperature[77]. The resulting product **41** precipitated as brown crystals when the pH was adjusted to neutrality. Barnett *et al.* published a method for the selective introduction of iodine in the 5-carbon position[62]: The *in situ*-silylation of both 4-oxygen and 7-nitrogen (see *scheme 17*) using bis-(trimethylsilyl)-acetamide should fix the heterocycle **41** in a defined electron configuration.

For pyrrolo[2,3-*d*]pyrimidines it is reported that the position of the electro-philic substitution strongly depends on the particular base substituents as well as on the reaction conditions[78]. If the 2-amino function were not protected by the pivaloyl group, it would direct the electrophilic attack of the iodine into the undesired 6-carbon position. The 6-iodinated product would be a result of the mesomeric stabilized σ-complex 1 shown in *scheme 17*. On the other hand, protection of the 2-amino function forms σ-complex 2, which would then favor the formation of the desired product. The silyl groups are labile enough for being removed quantitatively during the aqueous workup afterwards. Another important fact to mention here is the photosensitivity of NIS. If the reaction were carried out in the presence of light, a small amount of elemental iodine would be formed which could also lead to unselective iodination. Therefore we ran the iodination of heterocycle **41** in the absence of light, providing us with product **42** in moderate yield.

Unlike the synthetic steps before, the glycosylation of compound **42** employing Seela's method (see *scheme 16*) was not successful: Monitoring the glycosylation reaction via TLC showed that both the sugar as well as the heterocycle moiety did not react. The attempt to purify the crude reaction mixture on a short silica gel column delivered the heterocycle and the hydrolyzed sugar moiety, but not the desired nucleoside **43**. One reason for the inhibited glycosylation might be the competitive deprotonation of the three acidic protons of compound **42** (see protons in positions H2, H4 and H7 of compound **42** in *scheme 16*). Another effect may be that the iodine substituent in 5-carbon position lowers the nucleophilicity of the heterocycle and therefore hinders the nucleophilic attack on the chloro carbon on the sugar. As a result of this, no nucleophilic substitution would occur and the chloro sugar **27** would hydrolyze after 20 to 30 minutes if traces of water were present.

In order to enhance the reactivity of the heterocycle and to avoid unselective deprotonation it was decided to mask the 4-oxygen as chlorine function[65,76] (see scheme 18). The chlorination step readily converted compound **41** into the heterocycle **44**: The pivaloyl-protected heterocycle was refluxed for 2.5 h in a high excess of phosphorus oxychloride, then the excess was distilled off and the product **44** precipitated from aqueous ammonia[79]. The additional back-extraction of the aqueous layer furnished a smaller fraction of the chlorinated compound, too.

The selective 5-carbon iodination was accomplished again by using the Barnett method[62]: The best amount of product **45** was obtained when the iodination step was carried out in the absence of light. The iodinated product could also be isolated via precipitation with cold water and back-extraction. Both halogenations had to be carried out in the order chlorination first, then iodination. Once the reverse order was tried, but in that case it could be observed that the iodine in the 5-carbon position was partially removed during the chlorination.

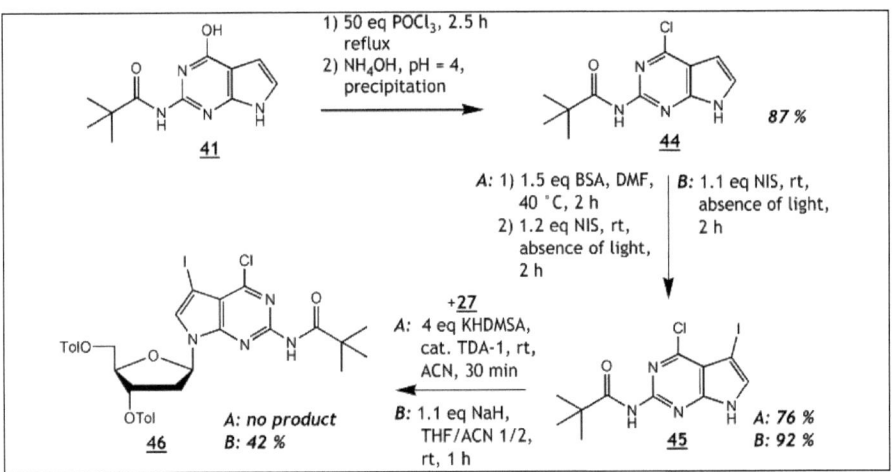

Scheme 18: Alternative method with masking the 4-hydroxy function

This was obvious when violet wads of gaseous iodine were generated during the refluxing of the reaction mixture consisting of phosphorus oxychloride and heterocycle **42**.

Due to the low solubility of compound **45** in acetonitrile causing low reactivity, the glycosylation following Seela's method still did not form the desired nucleoside **46**. Hence we changed the glycosylation conditions as follows: First, the solvent acetonitrile was exchanged by a mixture of THF and acetonitrile. Second, the more reactive base sodium hydride was applied. Both changes led to complete solvation and activation of the deazapurine moiety which finally reacted with the sugar ring to give the desired product **46**.

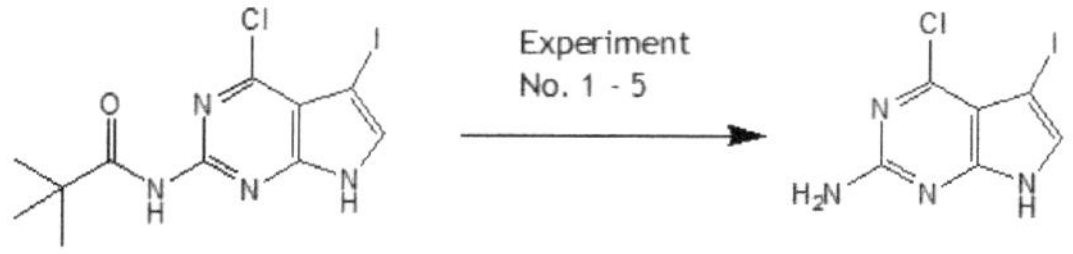

Exp. No.	Reagent	n [mol]	Solvent	T [°C]	Reaction time [h]	Yield
1	NaOMe[80]	1.1	MeOH	25	4	0
2	NaOMe[63]	2	MeOH	70 (reflux)	4	0
3	NaOH[77]	1	H_2O	25	72	0
4	HCl[81]	6	MeOH/H_2O	69	2	0
5	NH_4OH[76]	excess	-	100	3	0

Chart 1: Results of the pivaloyl deprotection attempts

The removal of the pivaloyl group on the 2-amino function using the unglycosylated heterocycle **45** caused unexpected problems: It was extremely stable under the conditions tested with and listed in *chart 1*. Although conditions 1 to 4 (see *chart 1*) are described in literature[63,76,77,80,81], the pivaloyl group could not be removed neither under alkaline nor under acidic conditions. Treatment with sodium methoxide in methanol did not convert the starting material into the deprotected product after stirring at room temperature for several hours[80]. Besides that, the 4-chlorine function of heterocycle **45** could be replaced by nucleophiles like water or methoxide during heating, which led to the following conclusion: If the cleavage conditions induce pivaloyl removal, the chlorine group is also affected – a fact that made this synthetic strategy useless for our purpose.

4.3.6 Synthesis of 2-amino-7-[2-deoxy-β-D-*erythro*-pentofuranosyl]-5-iodo-7*H*-pyrrolo[2,3-*d*]pyrimidin-4-one: Strategy 2

It was decided to drop the idea of using the pivaloyl protection for the 2-amino function and try a completely different synthetic approach, as shown in *scheme 19*. Here, both the 2-amino and the 4-carbonyl functions are masked[82]: The 4-carbonyl is replaced by chlorine and the 2-amino by a thiomethyl group so that no reactive protons, except the proton on 7-nitrogen, would be left.

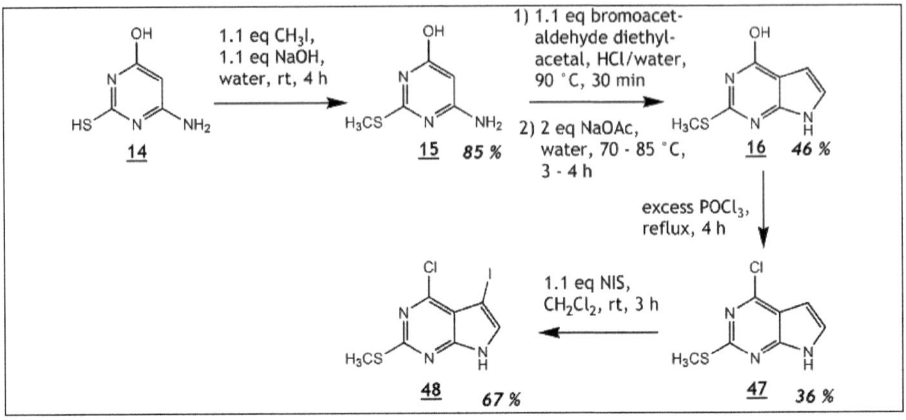

Scheme 19: Preparation of the masked heterocycle **48**

This circumstance should enable an efficient glycosylation with subsequent conversion of the masked groups into the original functionalities. The desired heterocycle **48** was accessible over four synthetic steps. The first three were described earlier for the synthesis of nucleoside **31**, but in this case the 2-methylthio group is not removed.

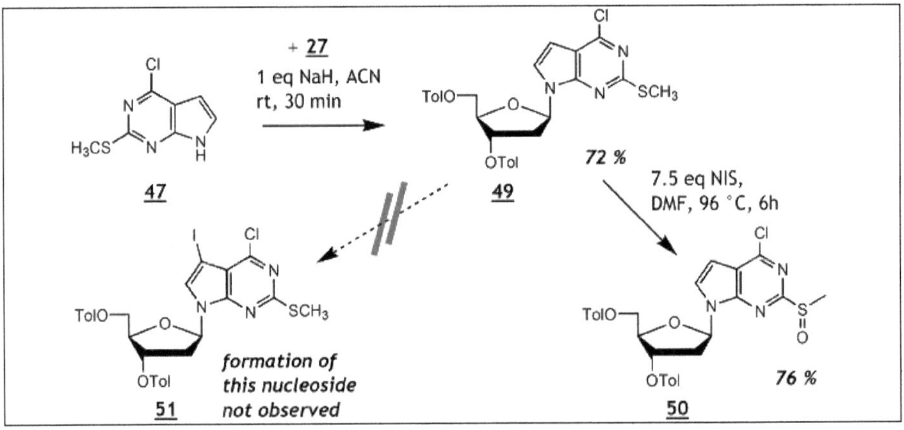

Scheme 20: The glycosylation with heterocycle **47** and subsequent iodination trial

Masking the 4-carbonyl function by refluxing heterocycle **16** in an excess of phosphorus oxychloride with subsequent quenching, aqueous workup and recrystallization gave compound **47**. Like it was already mentioned, the extraction and purification of the product worked poorly due to its low solubility which caused moderate yields around roughly 40 %. Compared to the chlorination step, the iodination in 5-carbon position using N-iodosuccinimide[62c] worked better and with high regioselectivity: In principle, the 2-methylthio function does not act as π-electron donor for the aromatic system, therefore the addition of iodine is not guided to 6-carbon, but to 5-carbon position. Performing the glycosylation[64] with sodium hydride (60 % in mineral oil) and heterocycles **47** or **48** in acetonitrile at room temperature led to the desired β-isomer **49**. Once it was also tried to introduce the iodine function after the glycosylation like it is mentioned in literature[82]: With the lack of the iodine function the heterocycle seems to be a slightly better nucleophile so that the glycosylation gave a higher yield (see *scheme 21*). The treatment of the resulting nucleoside with an excess N-iodosuccinimide in DMF at 96 °C led to an unexpected product. Instead of the formation of compound **51**, the corresponding oxidized nucleoside **50** was formed. This could be proven by ^{1}H-NMR and ESI(+)-mass spectroscopy as shown in *figure 18*. The mass of the expected product is 677.94, but in the ESI(+) spectra taken from the product two different main peaks occur. Compared to the mass of the oxidized compound (calculated 693.94), one signal can be found at 694.1 which belongs to the molpeak, another signal at 716.0 which corresponds to the sodium salt of the oxidized nucleoside.

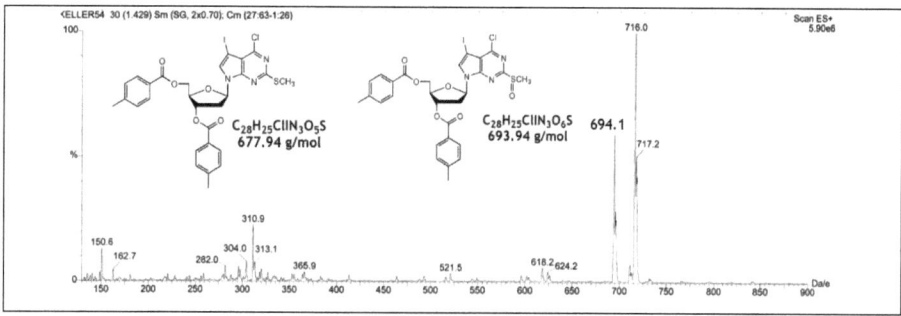

Figure 18: ESI(+)-MS spectrum of compound **50**

Finally the reconversion of the 2-amino and the 4-carbonyl function as shown in *scheme 21* delivered the desired nucleoside 2-amino-7-[2-deoxy-β-D-*erythro*-pentofuranosyl]-5-iodo-7H-pyrrolo[2,3-*d*]pyrimidin-4-one **54**: The chlorine in 4-carbon position of compound **51** could be replaced by using an excess of the oxygen-donating reagent *syn*-pyridine aldoxime in combination with the strong base 1,1,3,3-tetramethylguanidine: The aldoxime substututes the chlorine function supported by the base 1,1,3,3-

tetramethylguanidine. Quenching the reaction under acidic aqueous conditions always led to the desired product **52** in very good yield. For the replacement of the 2-methylthio function of nucleoside **52**, it had to be converted into a better leaving group first: oxidation of the sulfur atom with *m*CPBA led to the activated intermediate **53** which was purified on a short silica gel column. The activated species **53** was dissolved without further analysis in saturated ammonia/dioxane solution then, filled in a Parr bomb and heated at 110 °C for 15 h. Compared to what is reported in literature the yield of the fully substituted and deprotected nucleoside **54** was low.

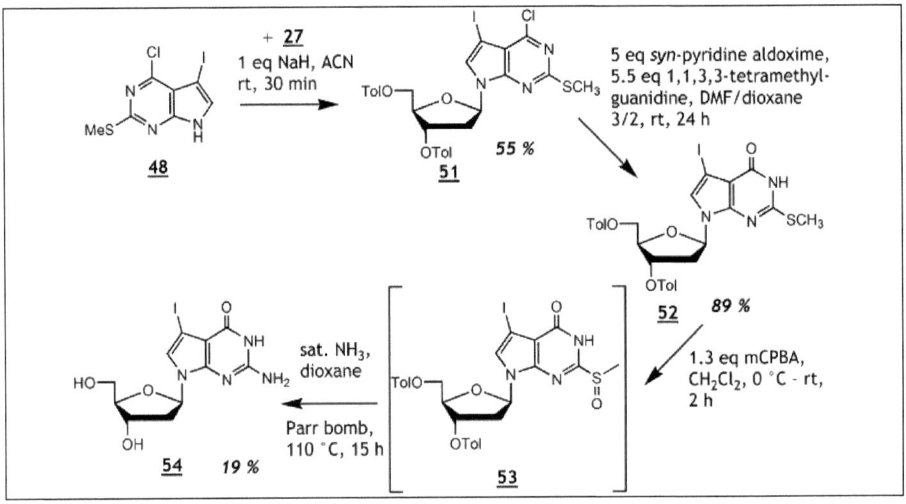

Scheme 21: Final steps and reconversion of both the 4-chloro and the 2-methylthio group[82]

The reason for this is that the conversion of the starting material led to some partially deprotected, partially substituted by-products during the reaction. Besides the low efficiency of the reaction some of the starting material also decomposed under these harsh conditions. Due to the time-demanding synthesis with poor overall yield (9 % over 4 steps) regarding the target compound **54**, we decided to acquire 2.0 grams of this compound which served as starting material for the preparation of the 3'-modified key compound.

4.3.7 Synthesis of 2-amino-5-iodo-7-[3-O-(2-cyanoethyl)-2-deoxy-β-D-*erythro*-pentofuranosyl]-7H-pyrrolo[2,3-d]pyrimidin-4-one

With 2 grams of 2-amino-7-[2-deoxy-β-D-*erythro*-pentofuranosyl]-5-iodo-7H-pyrrolo[2,3-d]pyrimidin-4-one **54** in hands we started with half amount of this material first to repeat the efficient protecting group strategy that was already evaluated by using 2'-deoxyguanosine. For the purpose of selective 3'-alkylation five steps for introduction of protecting groups were needed.

The first three steps are depicted in *scheme 22a*. The starting material **54** had to be well-dried for the introduction of the Markiewicz protection, therefore the nucleoside was coevaporated three times with pyridine and dried over two days in vacuum before use. Treatment of the nucleoside **54** with 1.1 equiv 1,1,3,3-tetraisopropyldichlordisiloxane in pyridine and stirring overnight at room temperature gave the protected nucleoside **55** in quantitative yield.

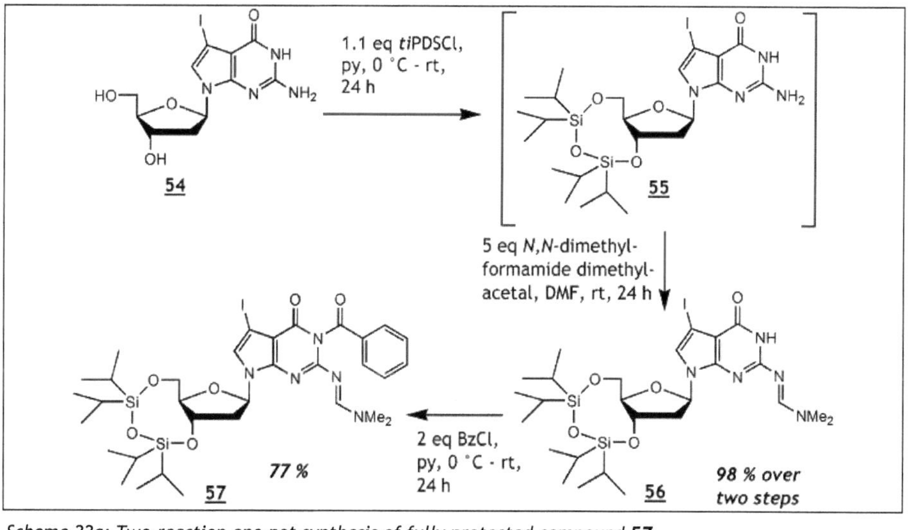

*Scheme 22a: Two-reaction one-pot synthesis of fully protected compound **57***

We did not purify compound **55**, but quenched the reaction by addition of methanol and the solvent removal under reduced pressure. We tried then to dissolve the oily residue in methanol, but unlike the Markiewicz-protected 2'-deoxyguanosine the nucleoside **55** did not dissolve. Therefore methanol was removed and DMF was added: Fortunately nucleoside **55** formed a clear solution then and was treated with an excess of N,N-dimethylformamide dimethylacetal. After stirring the mixture overnight at room temperature, product **56** was isolated after purification in excellent yield. The 3-nitrogen position was protected afterwards by treatment with 2 equiv benzoyl chloride (added in two portions) in pyridine in satisfying amount delivering the fully protected nucleoside **57**.

The 1,1,3,3-tetraisopropyldisiloxane moiety was cleaved off quantitatively by using an excess of triethylamine tri-hydrofluoride in THF (see *scheme 22b*).

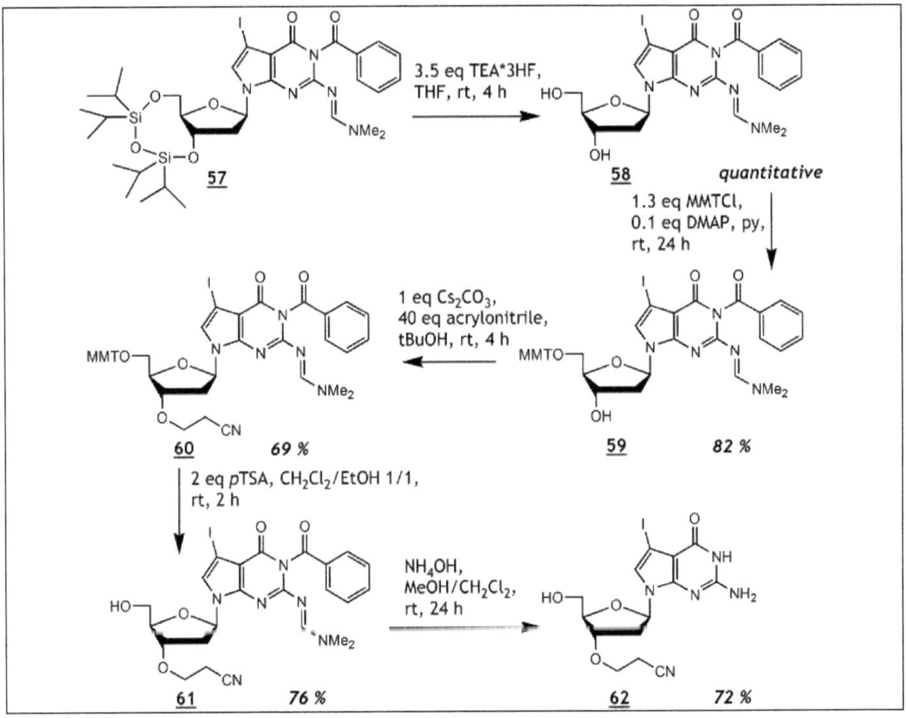

*Scheme 22b: The 3'-O-selective introduction of the CE function on compound **59***

After thoroughly drying of nucleoside **58**, the MMT-protection was carried out, catalyzed by the addition of 0.1 equiv of DMAP. The resulting compound **59** was now fully protected except its 3'-hydroxy group for enabling selective 3'-alkylation via the Michael addition[57a]. This time 40 equiv of acrylonitrile were needed for enhancing the solubility and conversion of the starting material. As a result of this we isolated the desired nucleoside **60** in a moderate yield of 69 %. With an excess of *para*-toluenesulfonic acid the MMT group was easily cleaved off to furnish compound **61** in 76 % yield. Both formamidine and benzoyl protecting group are known to be removable under alkaline conditions: By treatment of the nucleoside **61** with an excess of ammonia in water and methanol for 24 h, most of the starting material was converted into the deprotected nucleoside **62**. The minor product (less than 20 %), which has been characterized after aqueous workup and purification, was 3-O-[(2-cyanoethyl)-2-deoxy-β-D-erythro-pentofuranosyl]-2-(N,N-dimethylaminomethylidenyl)-5-iodo-7H-pyrrolo[2,3-d]pyrimidin-4-one. Based on this result we could conclude that the formamidino group on the 2-amino function is more stable than the benzoyl group on the 5'-end. In contrast to the other three 3'-O-(2-cyanoethyl) key compounds, nucleoside **62** was not modified further via Sonogashira reaction for the introduction of the dye-linker system [55b].

4.3.8 Synthesis of 5-[3-amino-prop-1-ynyl]-3'-O-(2-cyanoethyl)-2'-deoxycytidine

In contrast to the syntheses of both 3'-modified pyrrolo[2,3-d]pyrimidine nucleosides **38** and **62**, the third key compound **72** was prepared in a different way.

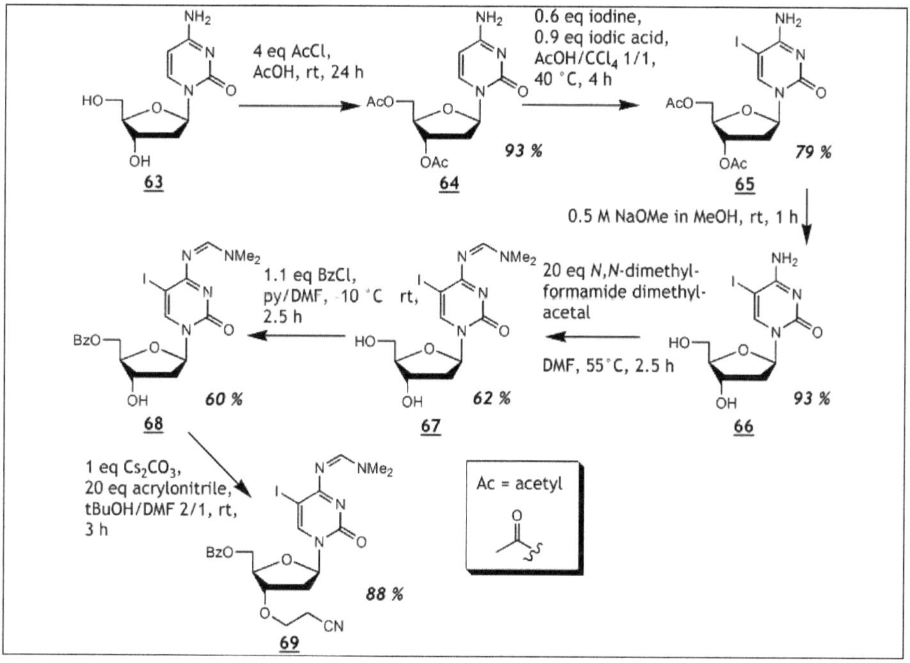

*Scheme 23a: Synthetic route for the preparation of nucleoside **69***

We started with commercially available 2'-deoxycytidine **63** and introduced the iodine function on the 5-carbon position. The desired nucleoside 5-iodo-2'-deoxycytidine **66** was accessible in moderate yield over three steps according to Bobek's method[83b] (see scheme 23a). The acetylation of both 3'-and 5'-hydroxyl functions worked in good yield[83a], although in our case the reaction often proved to be much more time-consuming due to the low solubility of the starting material. The iodination in the 5-carbon position proceeded highly selective: No diiodinated or 6-carbon-iodinated product was ever observed and isolated. The selective iodination can be explained by its mechanism as already claimed by Chang et al.[84] (see scheme 23b): They assume that two modes of iodination happen simultaneously, i. e. substitution and addition of iodine.

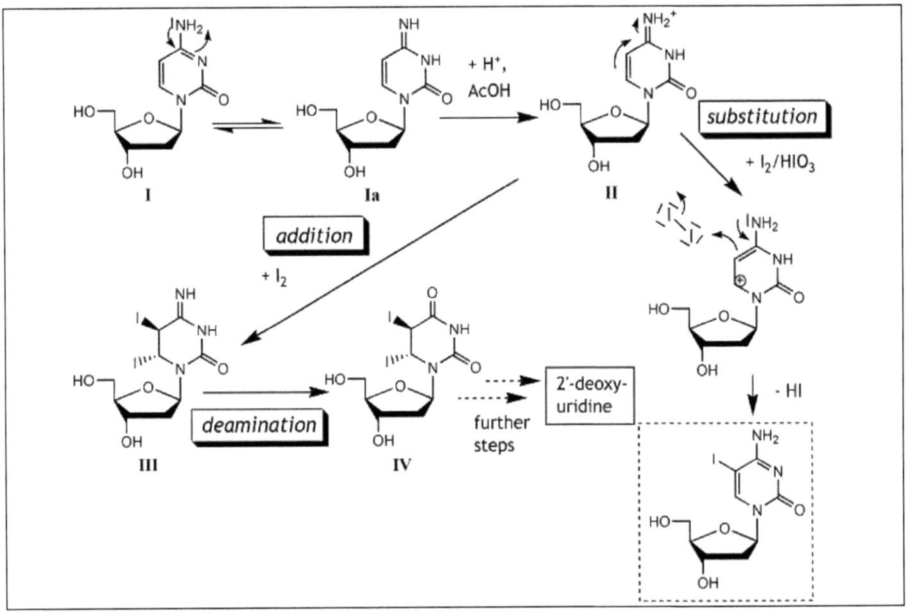

Scheme 23b: Proposed mechanism of the selective 5-carbon iodination by Chang et al.[84]

Carrying out the iodination with iodic acid[83b,84] as oxidizing agent favors the substitution and therefore the formation of the desired 5-carbon substituted product **65** without destroying the glycosidic bond. The substitution may be preferred because the 5,6-double bond of intermediate II (see scheme 23b) is more reactive in acidic medium favors substitution reaction. The addition reaction is less favored due to the fact that the imino group of intermediate III is more labile than before the addition of iodine.

Therefore 5,6-diiodo-2'-deoxycytidine IV is the minor product, which might lead to deamination with formation of 5,6-diiodo-2'-deoxyuridine and further synthetic steps[84]. After cleavage of the acetyl protecting groups the very polar and nearly insoluble product 5-iodo-2'-deoxycytidine **66** was obtained. Its low solubility caused only moderate yields in the next two steps, the formamidino protection of the 4-amino function[69] and the 5'-O-benzoylation[70]. The CE moiety was attached to the 3'-end by using the already described Michael reaction[57a]. When this reaction was carried out first, we observed that starting material **68** did only hardly dissolve in acrylonitrile and *tert*-butanol. Although 20 equiv of acrylonitrile and a few milliliters of *tert*-butanol were added extra, only roughly 50 % of the starting material was consumed after 3.5 h reaction time. After 4 h we stopped the reaction and isolated the product after purification on column causing cleavage of the protecting groups.

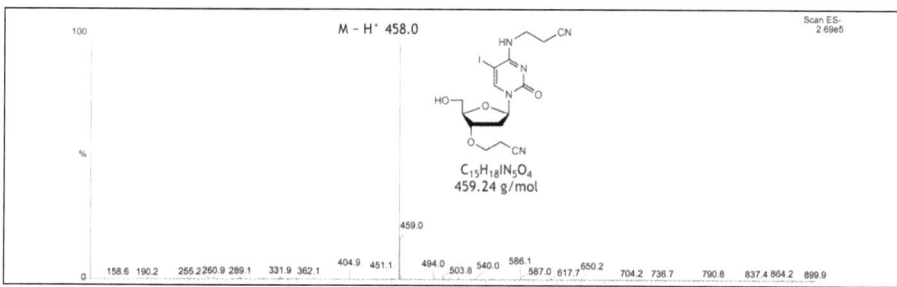

Figure 19: ESI(-)-mass spectrum of 2'-deoxy-(3',3N-dicyanoethyl)-5-iodocytidine

Thus we decided to completely cleave off both the formamidino[71] and the benzoyl[34a] protecting group in methanolic ammonia in order to have a pure compound for analysis. With ESI-mass spectrometry and ^{1}H-NMR spectroscopy it was found that not the desired nucleoside **69**, but the dialkylated compound 5-iodo-(3',3N-dicyanoethyl)-2'-deoxycytidine was formed (see mass spectrum in *figure 19*). The explanation for this may be that the formamidino group is only metastable and partially cleaved off during the Michael reaction. As a result of this, the exocyclic amino function may be cyanoethylated as well as the 3'-hydroxy function. In order to avoid the dialkylation, the solubility of the starting material was enhanced by addition of DMF as co-solvent (see *scheme 23a*). Besides that, we limited the reaction time to a maximum of 3 h and used only a few milliliters of *tert*-butanol/DMF. The purification of Michael adduct **69** was always complicated caused by the cleavage of the base-labile protecting groups during column chromatography. In some cases, we deprotected crude product **69** directly after the Michael reaction and characterized the nucleoside after purification. Both the formamidino and the benzoyl group could be removed simultaneously in saturated methanolic ammonia, giving 5-iodo-3'-O-(2-cyanoethyl)-2'-deoxycytidine **70** in moderate yield (see *scheme 23c*).

This compound was difficult to purify on silica gel column due to its high polarity caused by the free 5'-hydroxyl and the 4-amino function. As an alternative to column chromatography the product was crystallized from ethanol which caused lower yields, but delivered the product in higher purity. The Sonogashira reaction for the introduction of the propargylamine moiety was carried out with an excess triethylamine, TFA-protected propargylamine and a ratio of palladium(0)-catalyst/copper iodide 1/2 mol equiv in dry DMF as already described[73].

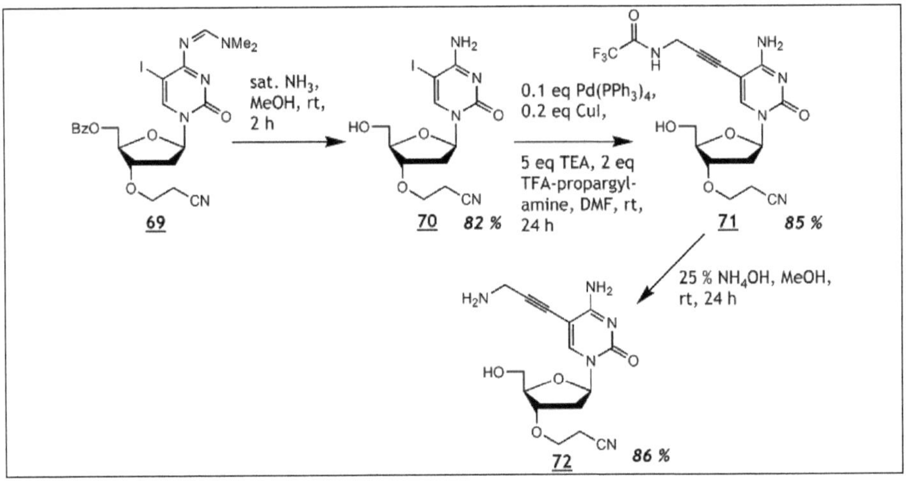

Scheme 23c: Final synthetic steps until the third key compound **72**

After concentration of the reaction mixture, product **71** was extracted with methylene chloride and washed with an aqueous 5 % disodium EDTA solution for removal of the metal cations. Concentration of the organic layer revealed only a minor product-containing fraction; most of it was still diluted in the aqueous layer. It was found that 2-butanone was an excellent extraction solvent for this highly polar product, which finally made it possible to extract the major fraction from the aqueous phase efficiently. The TFA group could also be removed under alkaline conditions[85] without formation of any by-products and finally key compound **72** was obtained.

4.3.9 Synthesis of 5-[3-amino-prop-1-ynyl]-3'-O-(2-cyanoethyl)-2'-deoxyuridine

The preparation of the key compound for the preparation of the T-terminator is comparable to the synthesis of nucleoside **72**. In this case, the starting material 5-iodo-2'-deoxyuridine **73** was commercially available so that no extra steps for the introduction of iodine into the 5-carbon position were necessary.

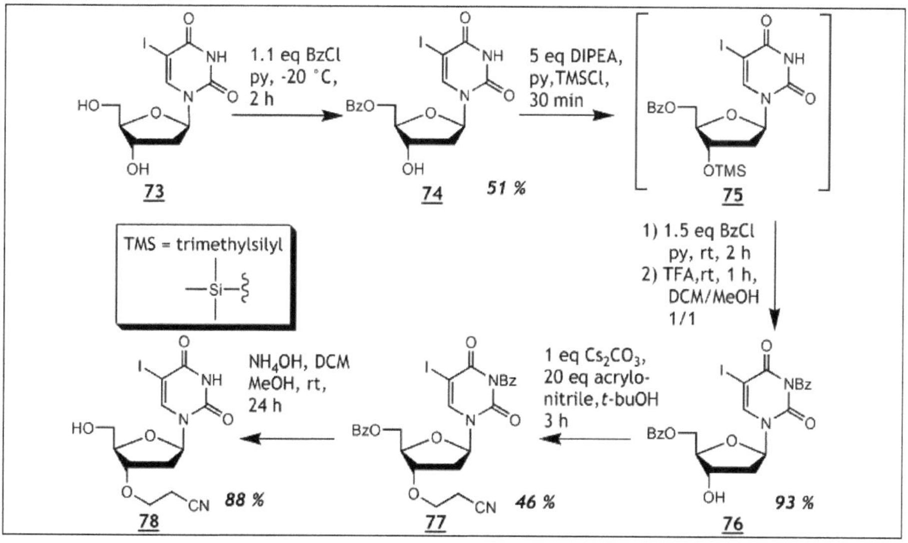

Scheme 24a: Synthetic strategy for enabling 3'-O-selective introduction of the CE function

As it is shown in *scheme 24a*, the 5'-hydroxy function was benzoylated in low selectivity: Apart from the expected product **74**, the corresponding 3'-benzoylated as well as the dibenzoylated nucleoside was formed. The selective N3-protection could be achieved by using the transient-protection method[86] employing trimethylchlorosilane as intermediate 3'-hydroxyl protection. After deblocking the 3'-hydroxyl function of the intermediate **75** with TFA, the dibenzoylated product **76** was obtained in high yield and purity. The CE function was introduced selectively on the 3'-position by using the already described Michael reaction[57a]. Here, the yield of 46 % product **77** after purification was moderate. As it is also the case for the cytidine analog, the benzoyl groups were partially cleaved off during purification of the product on column chromatography. Therefore reliable spectroscopic analysis was only possible after complete removal of both benzoyl groups.

The deprotected nucleoside **78** was formed by stirring the starting material **77** in a mixture of methanol and concentrated ammonia. The propargylamine-moiety (see scheme 24b) could be introduced via Sonogashira coupling[73] in satisfying yield: In this case the same ratio of palladium catalyst to copper iodide was employed (1/2 molequiv Pd-catalyst/CuI) like it was already used for the preparation of the key compound 5-[3-(trifluoroacetamido)-prop-1-ynyl]-3'-O-(2-cyanoethyl)-2'-deoxycytidine.

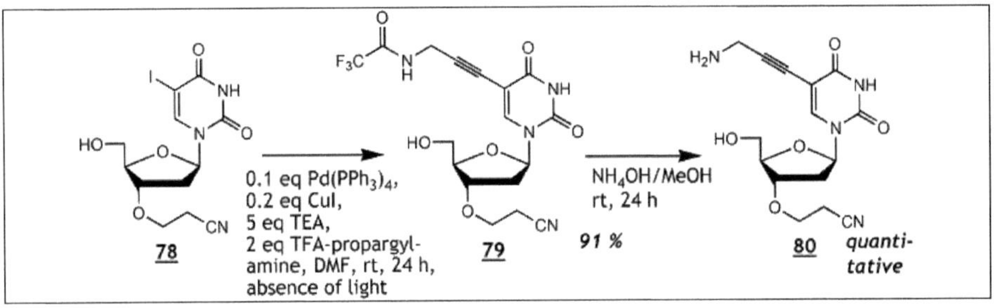

Scheme 24b: Final steps for obtaining the fourth key compound **80**

The trifluoroaceto group of nucleoside **79** was quantitatively removable in diluted ammonia/methanol delivering the key compound **80**.

5 Monophosphates as model compounds

5.1 Cleavage experiment in a heterogeneous system

As our reversible terminators possess a 3'-O-capping group like either the (2-cyanoethoxy)methyl (CEM) function or the 2-cyanoethyl (CE) function, it is of high interest studying and comparing the cleavability of both functions. There are two possibilities to check the cleavability of this group: One possibility is the incorporation of a terminal nucleotide bearing the 2-cyanoethyl function into an oligomer and subsequent cleavage experiments. As a simpler alternative to that the cleavage experiments can be carried out by using a stable and simple model compound, e.g. 3'-O-(2-cyanoethyl)-2'-deoxythymidine-5'-phosphate. In this case, the monophosphate moiety attached to the 5'-end lowers the solubility of the nucleotide in organic solvents like in THF or acetonitrile. This heterogenous environment simulates well the solubility of oligonucleotides in organic solvents enabling simple cleavage experiments without consumption of DNA-templates.

5.2 Synthesis of 3'-modified monophosphates

5.2.1 Synthesis of 3'-O-(2-cyanoethoxy)methyl-2'-deoxythymidine-5'-phosphate

The first two steps of the preparation of the CEM-wearing model compound are known in literature[34a] (see *scheme 26*). The Pummerer rearrangement[87] is the method of choice here for introducing the methylthio function on the 3'-hydroxy group. The mechanism[88] of this reaction, which can be regarded as inner redox reaction, is shown in *scheme 25*: Similar to ketones, the dimethylsulfoxide can tautomerize enabling the attack of its oxygen atom to the carbonyl function of acetic anhydride with release of one proton. An instable intermediate is formed then, which is stabilized after cleavage of the acetate moiety. The resulting compound is a sulfonium-ylide that can react in two different ways: Option 1 in *scheme 25* shows the possibility of a conjugated addition by an added alcohol to the electrophilic carbon of the sulfonium-ylide. With the nucleophilic attack of the hydroxyl function, another aliquot of acetate is released associating a proton in acidic environment (*i. e.* regeneration of acetic acid as catalyst). This reaction pathway leads to the methylthiomethylether as it is the case if compound **82** is used as substrate[34a].

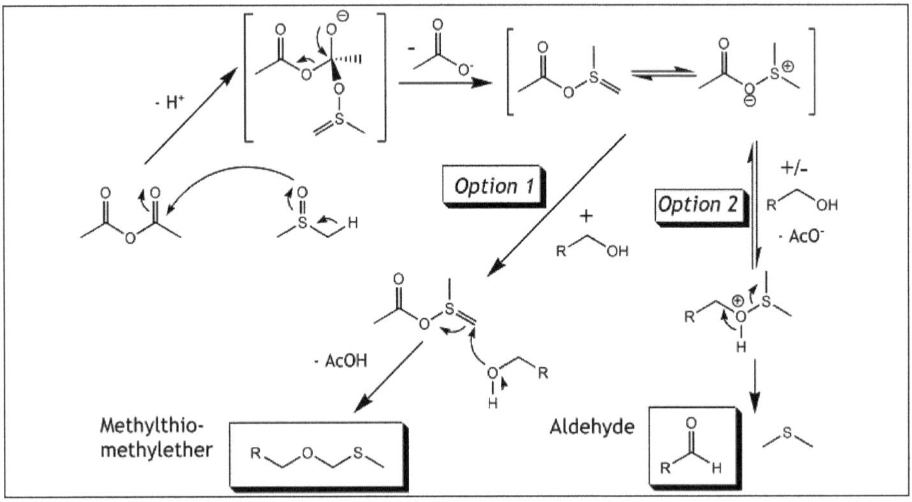

Scheme 25: Mechanism of the Pummerer rearrangement[88]

Another possibility (option 2 in *scheme 25*) is the nucleophilic attack of the alcohol's oxygen atom to the sulfur of the ylide forming equilibrium between acetate-bound and alcohol-bound ylide. The latter one of both intermediates is less stable and decomposes under oxidation of the alcohol to an aldehyde. This reaction pathway does not take place in our case, but might be dominant depending on the conditions and the nature of the alcohol[88,89].

As it is reported by Hovinen *et al.*, the reagents DMSO, acetic anhydride and acetic acid are not deployed in stoichiometric amounts but in excess and a defined volume ratio. With this simple procedure, stirring nucleoside **82** at room temperature for 24 h in this reaction mixture, the desired methylthiomethyl ether **83** could always be obtained in reliable yield (here 83 % as expected from literature[34a]). This nucleoside served as starting material for the introduction of the CEM function to the 3'-end. It is reported that the methylthiomethyl motif can be substituted with halogens like elementary bromine[34a], *N*-bromosuccinimide[34a,90] or, like in our case, with sulfuryl chloride[91] (see *scheme 26*). One has to be aware of the stability of intermediate **84**: Triethylamine as scavenger for hydrogen chloride, which was generated during the activation with sulfuryl chloride, prevented the decay of the activated chloro species.

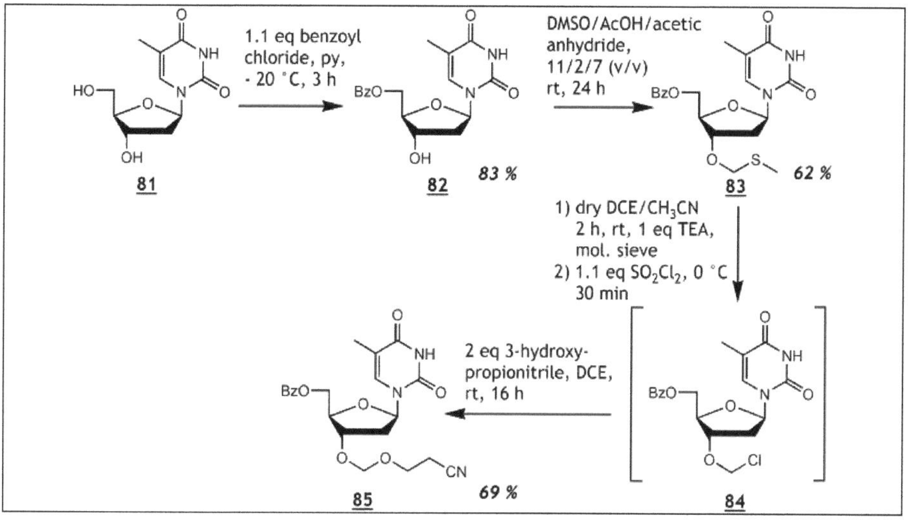

Scheme 26: Synthesis of 3'-O-CEM-dTMP part 1

Nucleophilic substitution of the chloro function with 3-hydroxypropionitrile in the next step led directly to formacetal nucleoside **85** in moderate yield. Here, also other nucleophiles could be added instead of this substrate giving a variety of different 3'-modified products[34a,90,91]. The removal of the 5'-protecting group gave nucleoside **86** ready for the final steps as shown in *scheme 27*. One big obstacle in obtaining desired nucleotide **88** was the metastable CEM group itself: By employing the Yoshikawa method[92] first, we did not isolate the product. Looking for alternative phosphorylation methods in order to get the desired monophosphate **88** in good yield, no convenient procedure was found in literature.

For this reason we decided to look for a published conventional triphosphate synthesis such as the Ludwig-Eckstein procedure[93] and modify it as a monophosphate synthesis. One possibility described by Gold *et al.*[94] and Sun *et al.*[95] is the two step procedure based on the Ludwig-Eckstein phosphorylation as first step with subsequent hydrolysis by addition of water as second step yielding quantitatively the H-phosphonate **87** (see *schemes 28a* and *28b*).

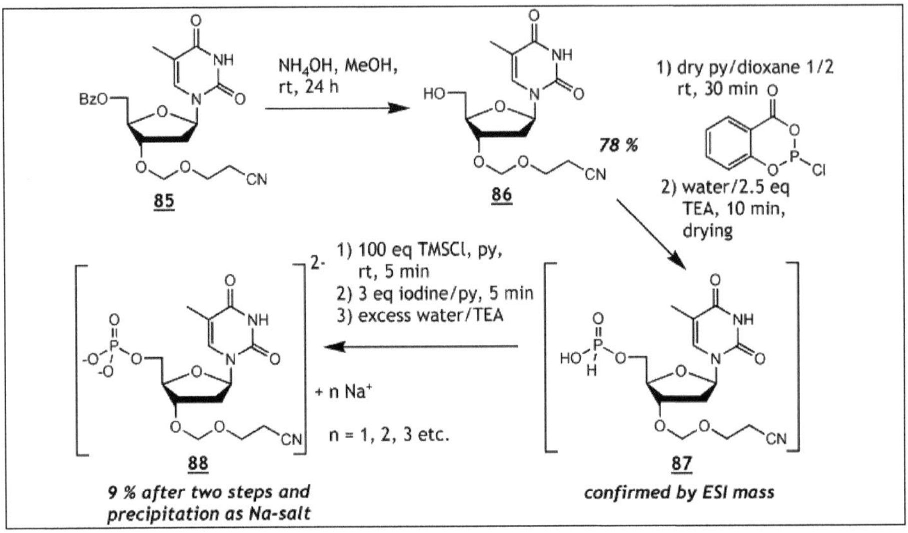

Scheme 27: Synthesis of 3'-O-CEM-dTMP part 2

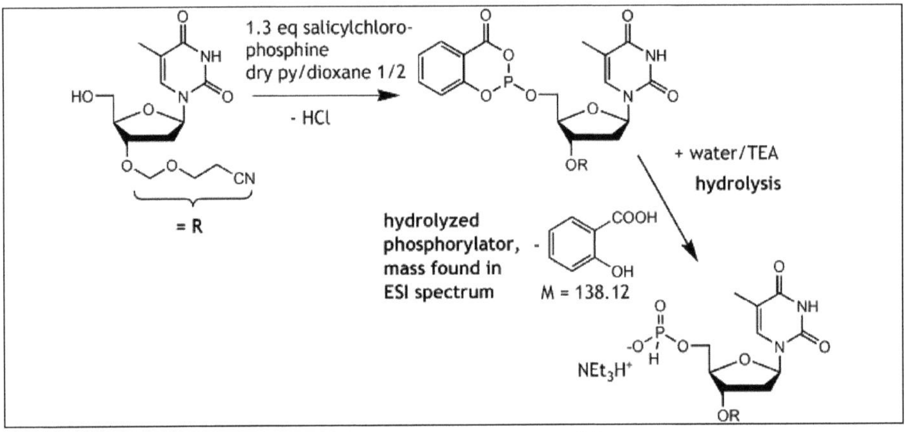

Scheme 28a: Mechanism of two-step phosphorylation method part 1[94]

This *H*-phosphonate is activated with an excess of trimethylchlorosilane in pyridine then facilitating the oxidation with iodine in pyridine. Subsequent addition of water would complete the formation of the phosphate (see *scheme 28b*). The proposed mechanism of this two step phosphorylation method is explained by Sun *et al.*[95]: The activation with TMSCl is claimed to convert the *H*-phosphonate I into a silyl-*H*-phosphonate IIa or more reactive bis-silyl phosphate IIb (see *scheme 28b*). The authors could prove the presence of similar intermediates like IIa and IIb because they analyzed the reaction mixture with mass spectroscopy and ^{31}P NMR: Significant changes in the chemical shift of the phosphorus atom after TMSCl addition as well as after iodine oxidation indicated that such intermediates must have been formed[95].

The subsequent addition of elemental iodine in pyridine would rapidly oxidize the P(III) species IIb *in situ* and generate the reactive pyridinium phosphor-amidate III (P(V) species)[96].

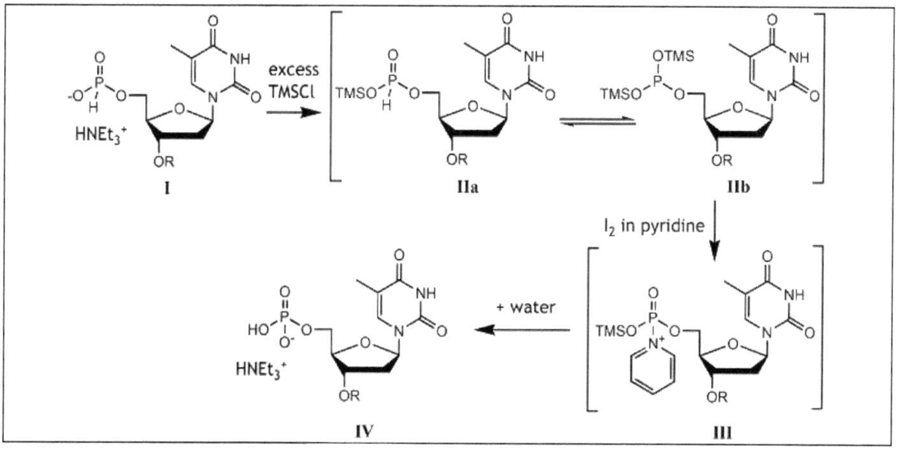

Scheme 28b: Mechanism of two-step phosphorylation method part 2 with proposed intermediates[94,95]

This intermediate could be transformed then into the desired phosphate IV by nucleophilic substitution with water. As shown in *scheme 27*, we applied this method which finally led to the desired model compound **88** in good purity, but low yield. Analysis with ESI-mass spectroscopy obtained after RP-FPLC and RP-HPLC purification showed that the obtained fractions consisted of dimers as well as higher phosphates. The minor product, sufficiently pure for CEM-cleavage tests, was precipitated as sodium salt for facile handling and storage. The ESI-mass spectrum of the *H*-phosphonate **87** formed as intermediate is shown in *figure 20*, the ^{31}P-NMR and ESI(-)-mass spectra of phosphate **88** are shown in *figures 21* and *22*.

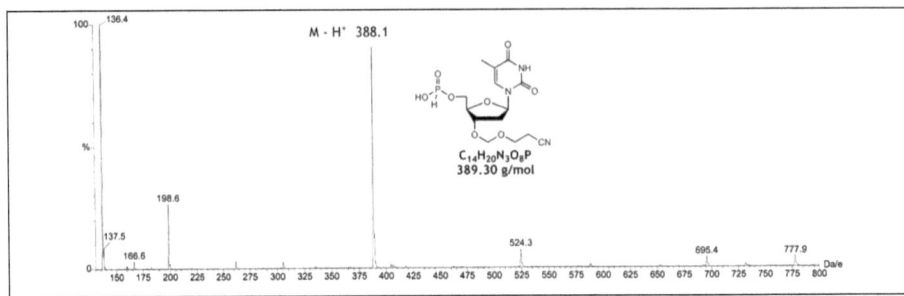

Figure 20: ESI(-)-mass spectrum of H-phosphonate **87**

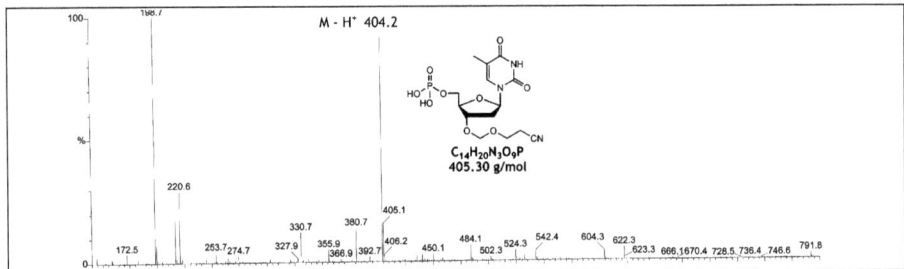

Figure 21: ESI(-)-mass spectrum of phosphate **88**

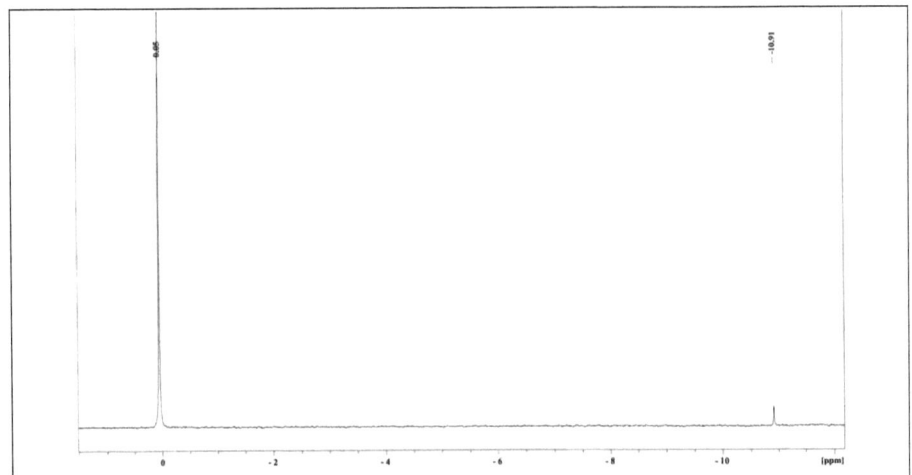

Figure 22: ^{31}P-NMR spectrum of phosphate **88**

In both mass spectra signals at 136.4/136.6 and 198.6/198.7 appear, which are in accordance with the calculated mass of the hydrolyzed phosphorylating agent 2-

hydroxybenzoic acid (136.6 g/mol) and its corresponding pyridinium-, Na⁺- and K⁺-salts. These impurities were not completely removable even with RP-HPLC, so it was decided to precipitate the monophosphate as sodium salt sufficiently pure for the deprotection tests.

5.2.2 Synthesis of 3'-O-(2-cyanoethyl)-2'-deoxythymidine-5'-phosphate

The synthesis starts with the already described one-flask transient protection of both hydroxyl-functions at the pentose-moiety of 2'-deoxythymidine (see *scheme 29a*). Following Sekine's procedure[86], 2.5 equiv TMSCl were added under alkaline conditions and after 30 min a slight excess of benzoyl chloride could be added to the reaction mixture. During the aqueous work-up, the silyl-protection was removed quantitatively and 3-nitrogen-protected nucleoside **90** was obtained in excellent yield after column chromatography. The tritylation of the 5'-hydroxyl group under standard conditions[57b,57c] led to nucleoside **91** which underwent a Michael addition[57a] reaction for introducing the CE function at the 3'-end (see *scheme 29a*).

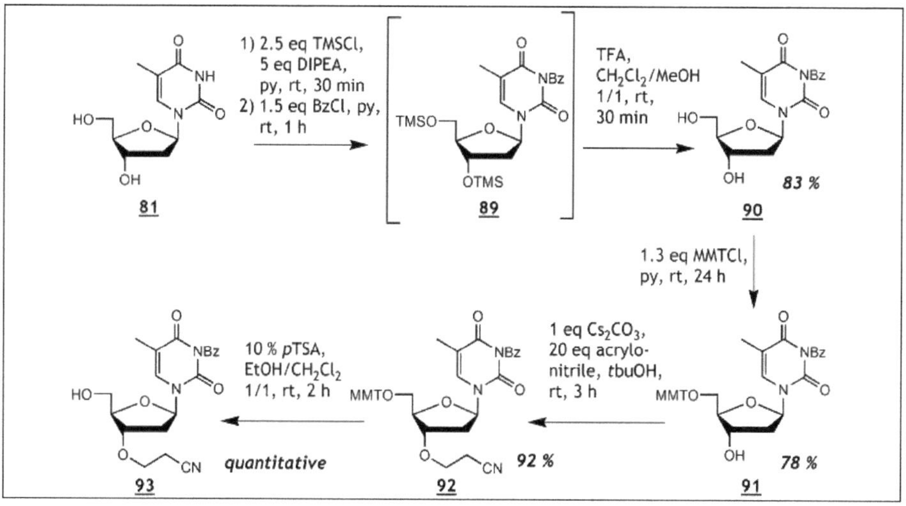

Scheme 29a: Protecting group strategy for 3'-O-selective introduction of the CE function

The MMT-protecting group of compound **92** could be removed quantitatively using *para*-toluenesulfonic acid. Without further purification, the crude product **93** could be concentrated under reduced pressure and dissolved again in methanol and aqueous ammonia (see *scheme 29b*).

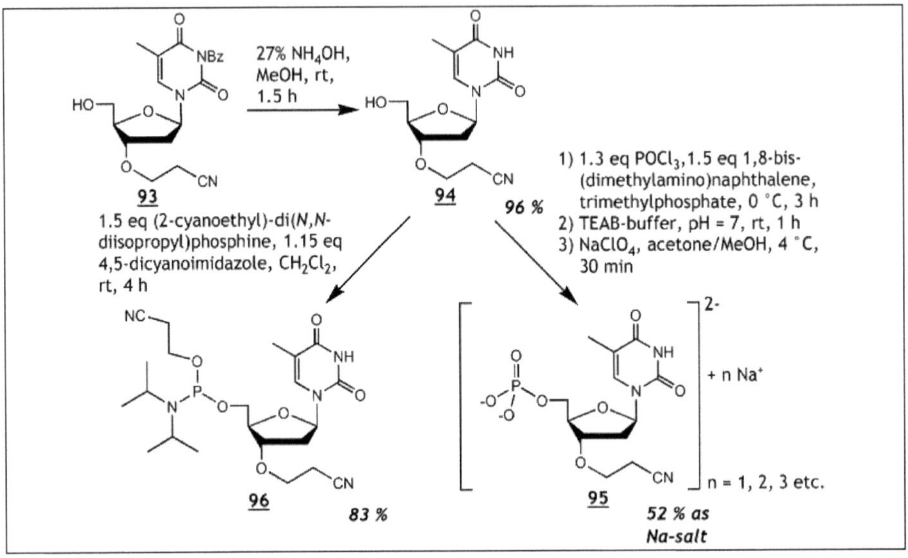

*Scheme 29b: Preparation of the monophosphate **95** and the phosphoramidite **96***

Stirring at room temperature yielded 5'-O-deprotected compound **94** without affecting the CE function, which means that this group was stable during both deprotection steps. The preparation of monophosphate **95** was simple, although the workup procedure appeared to be time-demanding due to product purification on semipreparative HPLC. We chose the conventional Yoshikawa method[92] for preparation of the monophosphate. The mechanism of this reaction is shown in *scheme 30*.

In literature it is described that the use of trialkyl phosphates as solvents shortens the reaction time of the phosphorylation[92,97]. This may be caused by the formation of a highly reactive "Vilsmeyer-Haack"-type complex made from phosphorus oxychloride and trimethyl phosphate which selectively phosphorylates the 5'-hydroxy group (see *scheme 30*). In some cases, 1,8-bis-(dimethylamino)naphthalene is used as proton sponge[98]. When we applied this method for conversion of nucleoside **94** into nucleotide **95**, we also added the proton sponge, but quenched the reaction with aqueous TEAB buffer in order to form the monophosphate.

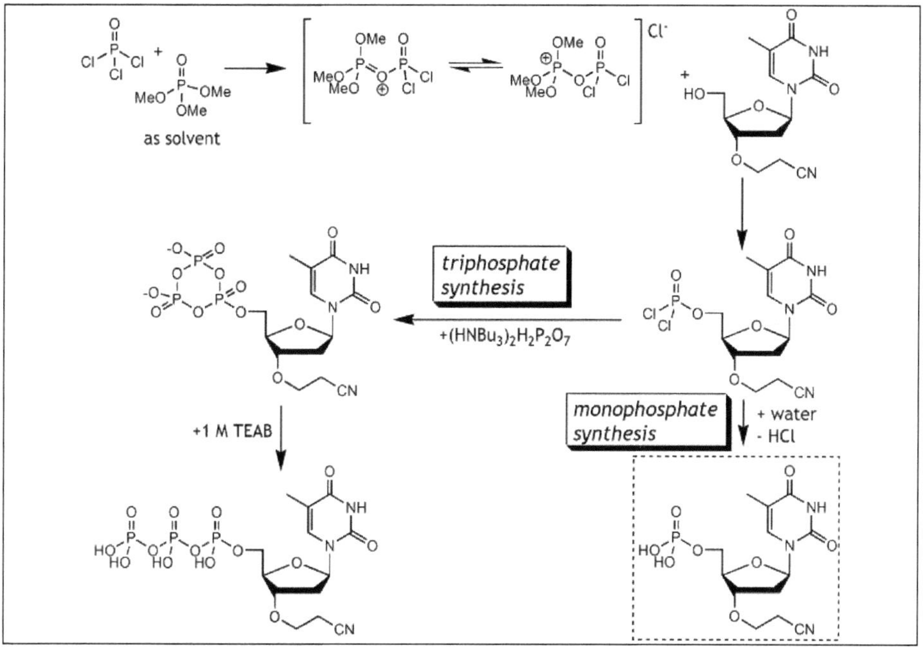

Scheme 30: Mechanism of the Yoshikawa phosphorylation[97,98]

The 5' hydroxyl function of nucleoside **94** was phosphorylated selectively in a moderate yield of roughly 52 % obtained after purification (see *scheme 29b*). Spectroscopic analysis via ^{31}P-NMR shows the singulett signal of the desired compound, but as well peak patterns that belong to a diphosphate and a triphosphate moiety.

As is mentioned in literature[98], a minor part of the product always contained those two by-products which could hardly be completely separated from the monophosphate on HPLC. We assume that the proton sponge also might support the formation of those di- and triphosphates built by repeated addition of reactive "Vilsmeyer-Haack" complexes to the 5'-hydroxy function of nucleoside **94**. The obtained monophosphate **95** was precipitated as sodium salt after RP- HPLC furnishing a colorless crystalline product, which is easier to handle and to store than glassy oil containing the anionic compound with undefined counter ions such as triethylammonium, magnesium or sodium cations in various ratios.

One fraction of nucleoside **94** was converted[69] into its corresponding phosphoramidite **96** for incorporation into a short DNA sequence (8 mer) for the CE deprotection test from an oligomer. The desired compound was obtained in moderate purity as can be seen from the ³¹P-NMR spectrum below (see *figure 23*): Two isomers of phosphoramidite **96** were formed and give one singulett signal each at 148.8 and 148.2 ppm, while the signal at 13.3 ppm indicates the presence of the H-phosphonate of compound **94**.

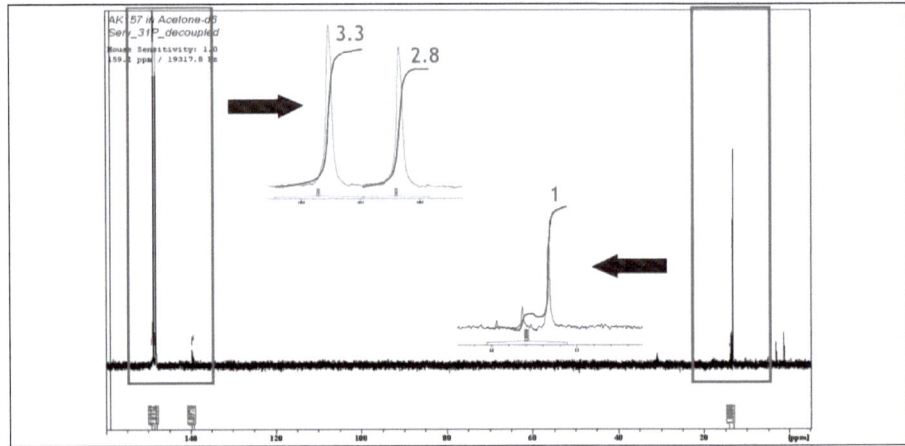

Figure 23: ³¹P-NMR spectrum of phosphoramidite **96** after purification

The integral ratio of both phosphoramidites together to H-phosphonate is 6/1, which means that the product consists of roughly 85 - 86 % of compound **96** and in 14 – 15 % amount of H-phosphonate and other impurities.

Besides the commercial availability of 2'-deoxythymidine-5'-phosphate, 2'-deoxythymidine **81** served as test compound for evaluation of efficient selective 5'-O-phosphorylation (see *scheme 31*). Applying the Yoshikawa method[92] without using the proton sponge not only proved to be the easiest way for obtaining the desired monophosphate **97** in moderate yield, but was also very selective with di-or triphosphates only present in traces (see ³¹P-NMR of compound **97**).

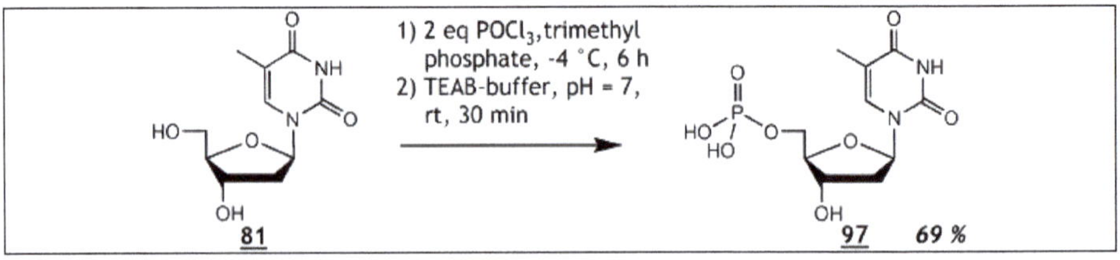

Scheme 31: Phosphorylation of 2'-deoxythymidine **81** as reference material

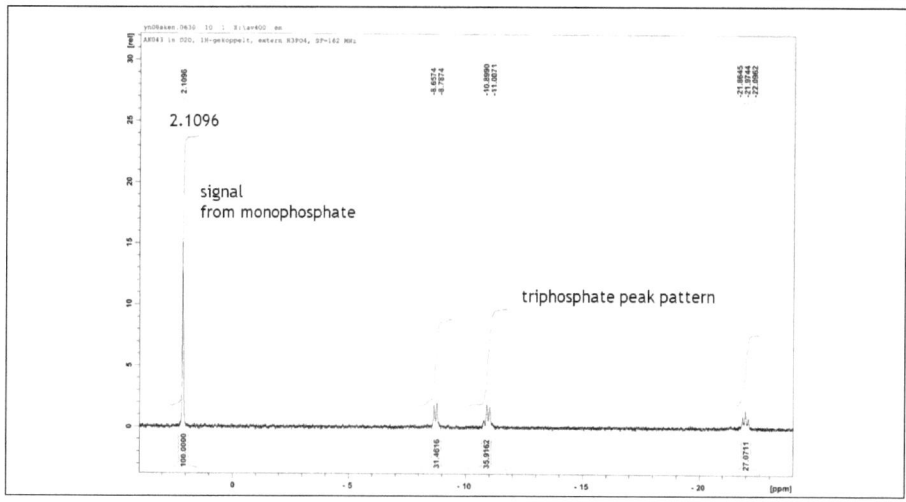

Figure 24: ^{31}P-NMR spectrum of nucleotide **95** after purification

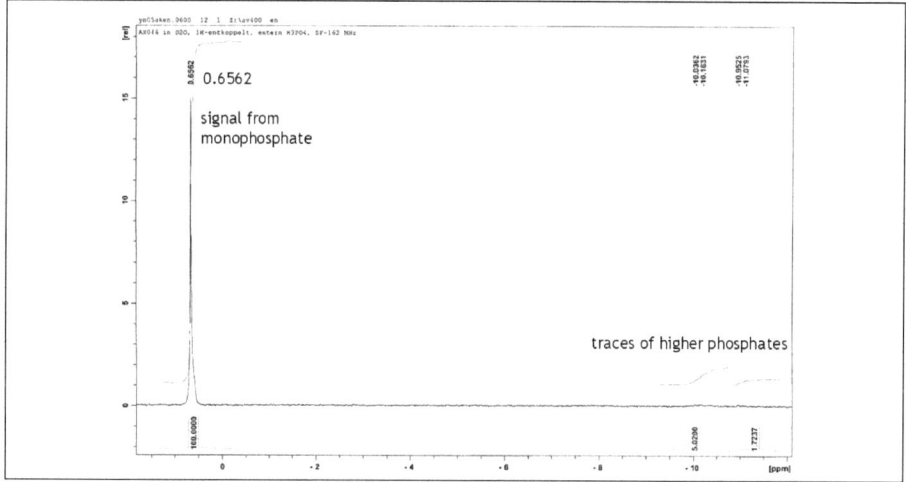

Figure 25: ^{31}P-NMR spectrum of nucleotide **97** after purification

The ^{31}P-NMR-spectra (see *figures 24* and *25*) of both nucleotides **95** and **97** demonstrate that the chemical shifts of the α-phosphorus from both compounds do not differ much. As can be seen from the ^{31}P-NMR spectrum of compound **95**, the Yoshikawa method carried out with a proton sponge favored the formation of the corresponding triphosphate as minor side product (roughly 20 - 25 % determined by signal integration). Carrying out the phosphorylation for the second time, 1,8-bis(dimethylamino)-

naphthalene was not added in order to avoid the formation of the by-product.

5.2.3   Synthesis of 3'-O-(2-cyanoethyl)-2'-deoxyadenosine-5'-phosphate

Starting with commercially available 2'-deoxyadenosine, the exocyclic amino function was protected first for preventing the side product formation in the next steps. Here, the formamidino protecting group seemed to be an optimal choice for selective and efficient protection[59a] (see *scheme 32a*). This reaction gave the desired product **99** in good yield after roughly 3 h reaction time with an excess of the formamidino acetal in DMF at 55 °C.

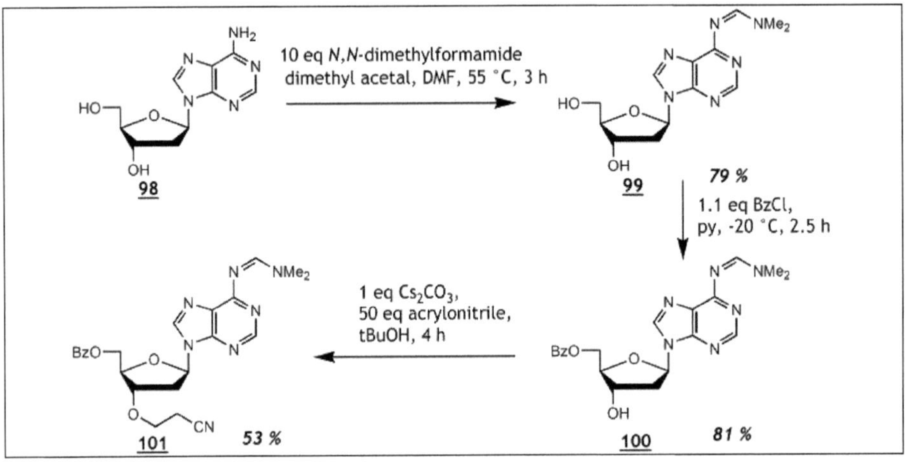

Scheme 32a: Synthesis of 3'-O-CE-dAMP part 1

The following benzoylation of the 5'-hydroxy function did not work selectively first: Employing methylene chloride as solvent for benzoyl chloride, the reaction delivered dibenzoylated product as main fraction and the monobenzoylated product **100** in 20 % yield only after purification. The yield increased significantly by using pyridine as single solvent and diluent for benzoyl chloride. In that case 81 % of the 5'-protected nucleoside **100** was achieved. After this step, the nucleoside underwent the Michael reaction performed at already known conditions, but with a higher excess of acrylonitrile (50 equiv) delivering 3'-O-cyanoethylated product **101** in moderate yield. After removal of the exocyclic protecting group as well as of the benzoyl moiety (see *scheme 32b*), the well-dried nucleoside **102** was converted into its corresponding nucleotide **104** by employing the two-step phosphorylation method[94,95] already used for the preparation of model compound **88**: The nucleoside was phosphorylated first and the intermediate quenched by addition of water, quantitatively yielding the H-phosphonate. The formation of H-phosphonate **103** was confirmed by ESI-mass spectroscopy as shown in its spectrum in *figure 26*.

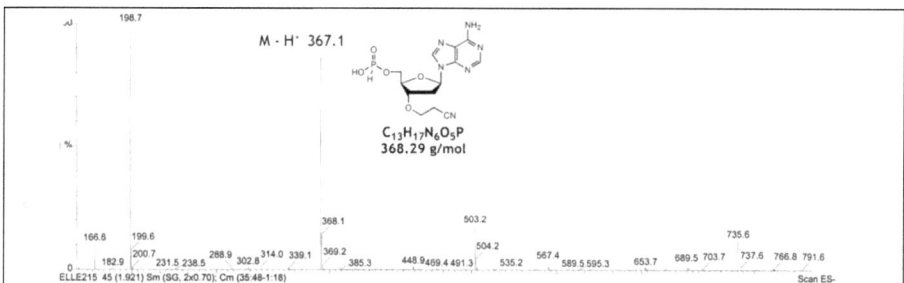

Figure 26: ESI(-)-mass spectrum of H-phosphonate **103**

This intermediate product was dried well and oxidized with iodine after activation with trimethylsilyl chloride, furnishing the desired monophosphate **104**.

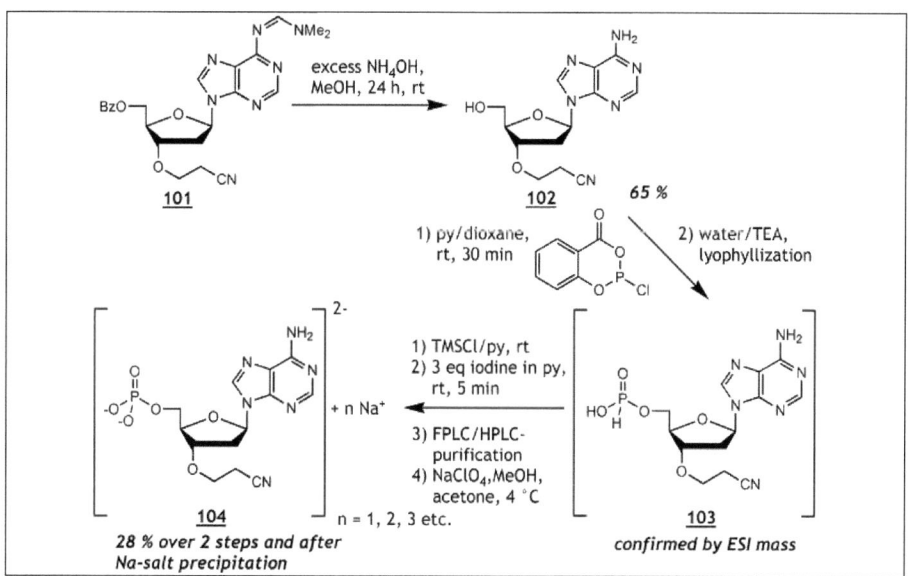

Scheme 32b: Synthesis of 3'-O-CE-dAMP part 2

The crude product **104** was purified in two steps then: First, the oily residue was taken up in a few milliliters of water and put on RP-FPLC after syringe filtration. In this purification step, excess of phosphorylator, inorganic salts and rests of starting material were separated from the product-containing fraction. The product was purified further on RP-HPLC in order to remove impurities and higher phosphates like dimers etc. as found by ESI-mass spectroscopy.

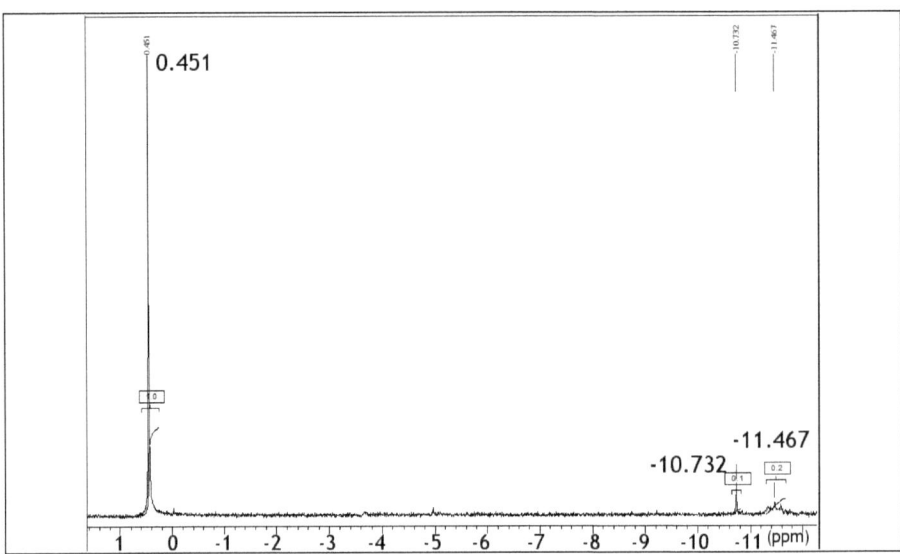

Figure 27: ^{31}P-NMR spectrum of compound **104**

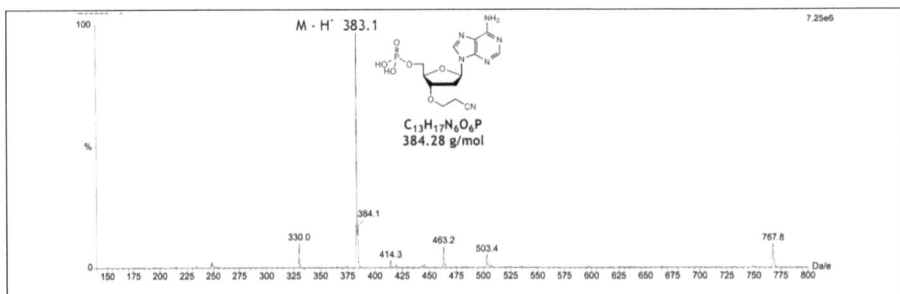

Figure 28: ESI(-)-mass spectrum of compound **104**

Both ^{31}P-NMR and mass spectra taken after RP-HPLC purification show impurities from the phosphorylating agent as well as signals from other undefined phosphorus species (see *figures 27* and *28*). Nevertheless the purity of phosphate **104** was sufficient for a few CE cleavage tests. For this purpose, the nucleotide could easily be stored and handled after its precipitation as sodium salt.

5.3 Cleavability of the 3'-modified monophosphates

5.3.1 Cleavage of the CEM function using 3'-*O*-CEM-dTMP

The cleavage tests performed with all three model compounds were carried out in the same experimental setup as shown in *figure 29*. This setup enables tempering and stirring of the reaction mixture, also samples can be taken out easily. Before its use, the apparatus was well-dried with heating in vacuum and flushed with argon in order to create a water-free environment.

Figure 29: Deprotection experiment setup with detailed view on reaction vessel

For each cleavage experiment, a certain amount of monophosphate **88**, **95** or **104** as sodium salt (in case of 3'-*O*-CEM-dTMP 1 mg) was suspended under argon in THF, acetonitrile or any other solvent (0.5 - 1 ml) and heated up to a given temperature. Treatment with one of the various cleavage reagents and taking out aliquots (80 µl) from the reaction mixture after certain time periods enabled monitoring of the reaction progress. These samples were quenched on demineralized water (120 µl) and injected into RP-HPLC directly. The detailed method and buffers are listed in the experimental part. As reference material for the cleavage tests done with model compound **88**, commercially available 2'-deoxythymidine-5'-phosphate **97** was coinjected (see *figure 30*).

The resulting chromatogram shown in *figure 30* displays both signals and retention times of 2'-deoxythymidine-5'-phosphate **97** (7.93 min) and 3'-*O*-(2-cyanoethoxy)methyl-2'-deoxythymidine-5'-phosphate **88** (11.64 min). It is obvious that both compounds are well distinguishable and allow monitoring of the CEM cleavage reaction.

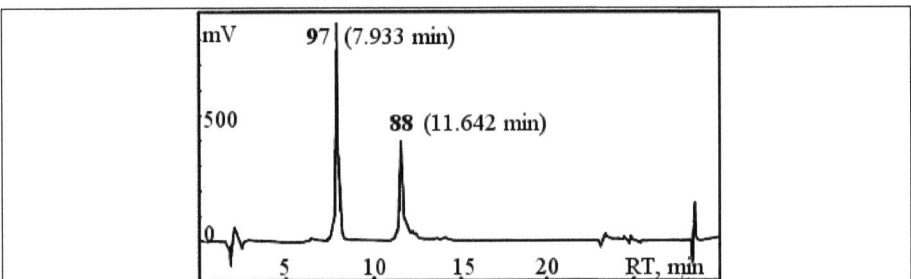

Figure 30: RP-HPLC chromatogram of model compound 88 and reference phosphate 97

The results of all deprotection experiments using model compound **88** are shown in chart 2.

Exper. No.	Reagent(s)/ solvent	Equiv. [mol-%]	Temp. [°C]	Time [min] for quantitative CE-cleavage
T_CEM_1	1 M TBAF/THF	40	40	6
T_CEM_2	0.5 M AcOH/water	~87000	60	No cleavage
T_CEM_3	TFA/acetonitrile	200	60	No cleavage
T_CEM_4	0.2 M KOH/water	40	60	35
T_CEM_5	TEA*3HF/ acetonitrile	10	60	No cleavage

Chart 2: Conditions for CEM-cleavage experiments using monophosphate 88

The CEM group is known as stable 2'-protecting group for RNA-synthesis[99] and 2'-*O*-CEM-derivatized ribonucleotides are reported to be cleavage with 0.1 M acetic acid at 90 °C[100]. In case of employing our model compound **88** possessing the CEM function as 3'-protecting group we obtained an unexpected result:

The CEM group was found to be remarkably stable under acidic conditions, while it was removed under alkaline conditions: Treatment with high excess of acetic acid (pH 2) at 60 °C neither led to CEM removal nor to nucleotide decomposition (see stacked chromatograms in *figure 31*). In contrast to that, dissolving the nucleotide in aqueous potassium hydroxide led to cleavage of the CEM group (see stacked chromatograms in *figure 32*). In the heterogeneous system nucleotide/TBAF/THF, the substrate was only suspended in THF, but cleavage of the CEM group was even more effective than before.

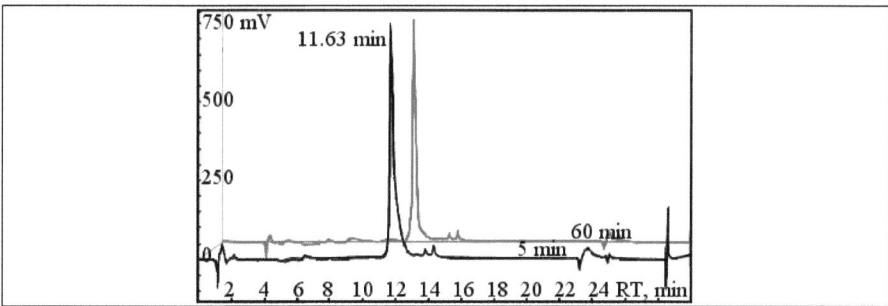

Figure 31: Stacked RP-HPLC chromatograms of cleavage experiment **T_CEM_2**

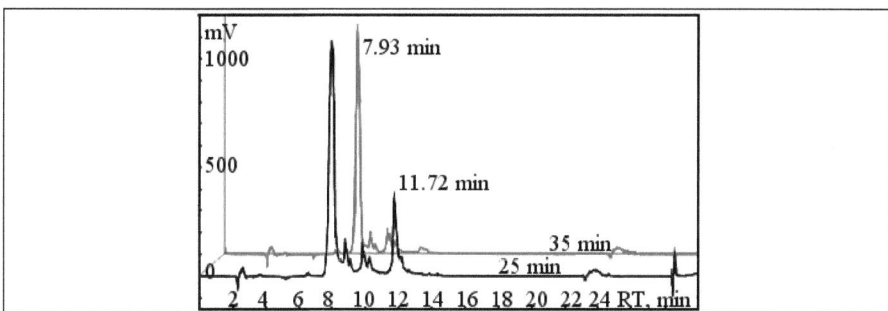

Figure 32: Stacked RP-HPLC chromatograms of cleavage experiment **T_CEM_4**

From these results we made the following conclusion: The dependency of the CEM removal on alkaline environment is caused by the mechanism of the cleavage, which is similar to the mechanism for the 1-(2-cyanoethoxy)ethyl (CEE) cleavage[101]: The removal of the CEM group is supposed to be a β-H-elimination as shown in *scheme 33*. Only small bases that can abstract β-hydrogens induce CEM decomposition and its removal.

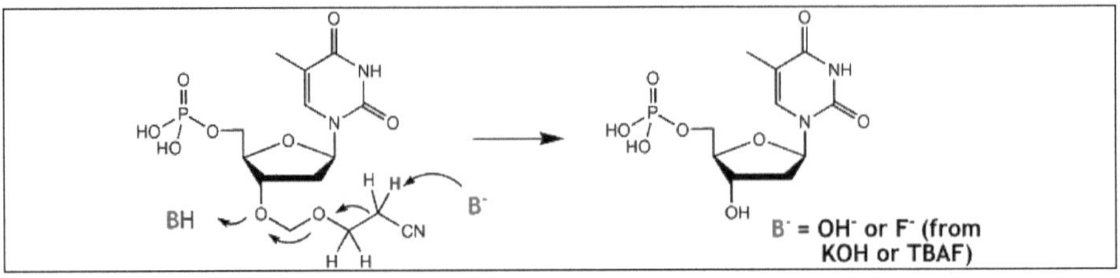

Scheme 33: Proposed mechanism of the CEM cleavage

Under acidic conditions, this reaction cannot take place, but either in heterogeneous or homogeneous alkaline environment. We also assume that the smaller and more electronegative the base is, the more effective the elimination reaction is. We already published the results from the first CEM cleavage tests made on an oligomer which is in accord with the results from the CEM cleavage tests of the monomer **88**[102].

5.3.2 Cleavage of the CE function using 3'-O-CE-dTMP

As we were even more interested in the behavior of the CE group, we investigated further in CE- than in CEM- cleavage tests. Thus we synthesized monophosphates **95** and **97** which were used as reference materials for the following cleavage experiments. The RP-HPLC chromatogram of both starting material (model compound) and deprotected monophosphate are shown in *figure 33*, while the commercially available phosphate **97** was coinjected for calibration of the method (see chromatogram in the background of *figure 33*).

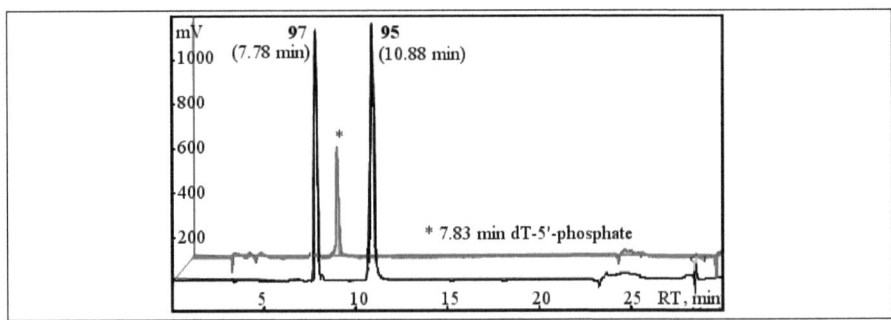

*Figure 33: RP-HPLC chromatogram of model compound **95** and reference phosphate **97***

The stacked chromatograms in *figure 33* show that the model compound after CE cleavage, which is assumed to be 2'-deoxythymidine-5'-phosphate **97**, is well separable from the starting material by the RP-HPLC method we have developed. The deprotected nucleotide **97** elutes significantly (at 7.78 min) earlier from the column than the CE-protected nucleotide **95** (10.88 min). This separation method was also used for the analysis of the cleavage experiment probes.

Dry 3'-O-(2-cyanoethyl)-2'-deoxythymidine-5'-O-phosphate **95** (often 20 mg) was suspended in dry solvent and heated up to the desired reaction temperature with stirring. After tempering the suspension for a few minutes on that temperature, the cleaving reagent was added in one lot and the stop watch started. After certain time periods, a sample (100 or 200 µl) was taken out of the reaction chamber and quenched on 500 µl demineralized water. These probes were filled after syringe-filtering into small vials and given to analytical HPLC via autosampler. The results of all deprotection experiments using the model compound **95** are shown in scheme (chart 3).

Exper. No.	Reagent(s)/ solvent	Equiv. [mol-%]	Temp. [°C]	Time [min] for quant. CE-cleavage
T_CE_1	1 M TBAF/THF	40	25	60
T_CE_2	1 M TBAF/THF	80	37	60
T_CE_3	1 M TBAF/THF	80	66	3
T_CE_4	1 M TBAF/THF	40	60	6
T_CE_5	TEA*3HF/ acetonitrile	10	60	No cleavage
T_CE_6	1 M TBAF/THF	10000	60	Not determined
T_CE_7	1 M solid TBAF/THF	40	60	10
T_CE_8	KOH/water	40	60	30
T_CE_9	Dicyclohexano-18-crown-6/KF/THF	.1 crown ether KF	60	No cleavage
T_CE_10	1 M TBAF/THF	5	60	>20
T_CE_11	1 M TBAF/THF	20	60	10
T_CE_12	1 M TBAF/THF/DMF (THF/DMF 1/1)	40	60	1.5
T_CE_13	1 M TBAF/THF/DMSO (THF/DMSO 1/1)	40	60	1
T_CE_14	1 M TBAF/THF/DMF (THF/DMF 1/1)	40	40	12
T_CE_15	1 M TBAF/THF/DMSO (THF/DMSO 1/1)	40	40	3
T_CE_16	TASF/DMF	15	25	No cleavage

Chart 3: Conditions for CE-cleavage experiments using model compound **95**

No cleavage was observed in acidic environment as can be seen from experiment no. T_CE_5 in chart 3. The stacked chromatograms of this experiment are shown in figure 34 revealing that the CE group is stable under acidic conditions and not cleavable with the acidic fluoride species hydrogen fluoride. The CE deprotection was also not induced by employing the phase-transfer catalyst 18-crown-6-dicyclohexano ether in combination with potassium fluoride as base[103]. These results support the assumption that the CE cleavage follows a β-H-elimination mechanism similar to the CEM cleavage mechanism, which is strongly dependent on the pH value of the reaction, the temperature and the nature and amount of base (see *scheme 34*). The elimination products acrylonitrile and the 3'-deprotected nucleotide **97** are still reactive if the reaction is not quenched properly and base B⁻ is still present in the reaction mixture. The reversibility of the elimination became more obvious when we carried out the cleavage tests with the purine analog **104** of the model compound **97** (see next chapter).

We also carried out one deprotection test with the fluoride species tris(dimethylamino)sulfonium difluorotrimethylsilicate (*i. e.* TASF, entry no. T_CE_16 in chart 3) which is known as a mild reagent for silicon protecting group removal[104].

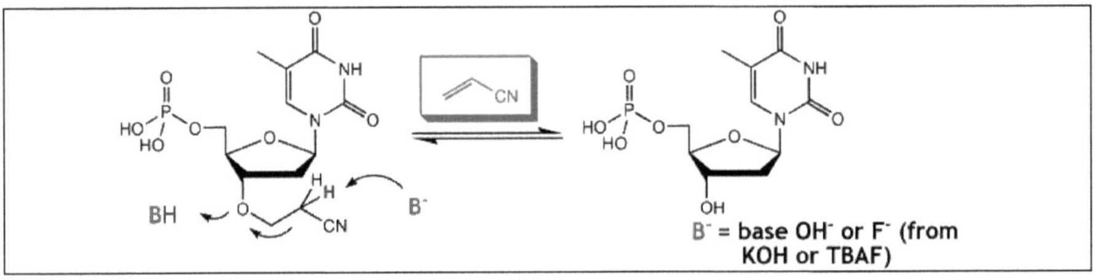

Scheme 34: Proposed mechanism of the CE cleavage

This reagent could also not remove the 2-cyanoethyl function, so finally we came to the following conclusion: The CE function is cleaved in an alkaline environment with strong and charged bases like fluoride; the cleavage rate is significantly improved at elevated temperatures, for example at 60 °C. Furthermore, the cleavage efficiency also depends on the solvent and TBAF amount.

As a result of deprotection tests, the cleavage in the model compound is enhanced by addition of polar solvents like DMF or DMSO to 1 M TBAF/THF standard solution (see entries no. T_CE_12, T_CE_13, T_CE_14 and T_CE_15 in *chart 3*). The stacked chromatograms shown in *figure 35* illustrate the enhanced CE cleavage by addition of DMF to the TBAF/THF-reaction mixture. After 3 minutes, the heterogeneous solvent/substrate system turns into a nearly homogeneous mixture and most of the starting material is consumed.

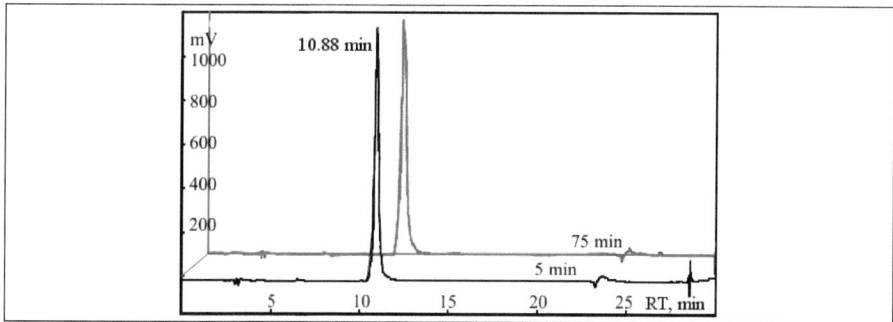

Figure 34: Stacked RP-HPLC chromatograms of CE cleavage experiment T_CE_5

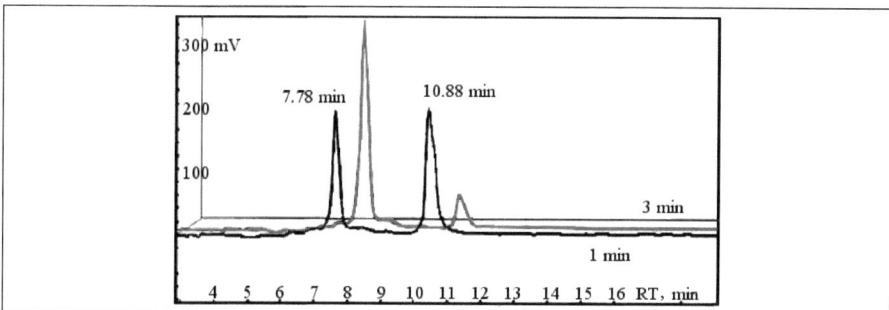

Figure 35: Stacked RP-HPLC chromatograms from CE-cleavage experiment T_CE_14

Based on these promising results from the effective CE cleavage by using monophosphates as model compounds, we investigated further in the CE cleavage from an oligomer. Due to the results from the polymerase acceptance tests, where 3'-O-CE-dTTP revealed better incorporation properties than 3'-O-CEM-dTTP, we focused only on the CE cleavage.

5.3.3   Cleavage of the cyanoethyl function using 3'-O-CE-dAMP

After evaluation of quantitative CE cleavage using the pyrimidine model compound **97**, we transferred our knowledge gained from these cleavage tests to the CE cleavage tests employing the purine model compound 3'-O-(2-cyanoethyl)-2'-deoxyadenosine-5'-phosphate **104**.

The cleavage reactions were carried out in the same apparatus as described before. For each cleavage test, a certain amount of model compound **104** (0.5 or 1 mg) was suspended in a given solvent and heated up to the desired temperature. The cleaving agent was added and after certain time periods, aliquots (80 µl) were taken out, quenched on demineralized water (120 µl) and injected directly into analytical RP-HPLC (method and buffers described in detail in the experimental section).

The chromatogram of both model compound **104** and commercially available 2'-deoxy-adenosine-5'-phosphate (**dAMP**), which is supposed to be identical with the model compound after CE deblocking, is shown in *figure 36*.

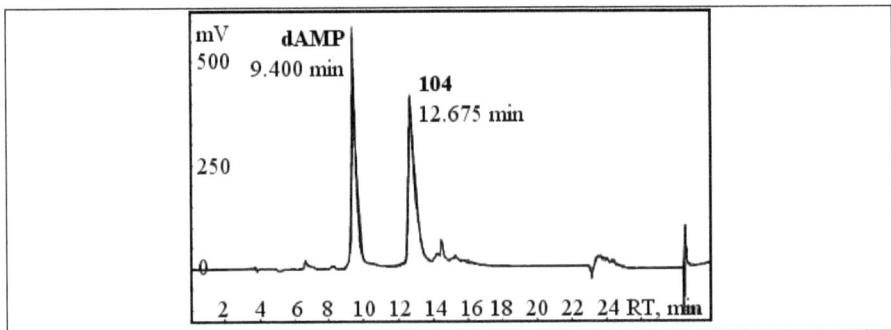

*Figure 36: RP-HPLC chromatogram of model compound **104** and reference phosphate **dAMP***

Here, both nucleotides are also well distinguishable: The deprotected nucleotide **dAMP** elutes significantly earlier (at 9.4 min) than the CE-protected nucleotide **104** (at 12.68 min). The results from the CE-cleavage experiments are summarized in *chart 4*. The first two experiments A_CE_1 and A_CE_2 didn't reveal any unexpected results: As we already know, the CE-function is cleavable with TBAF in THF by using minimum 40 equiv. TBAF at 40 °C for roughly complete CE-removal (*chart 4*, experiment no. A_CE_1).

In contrast to that, the CE function is stable under acidic conditions as experiment no. A_CE_2 shows: An excess of acetic acid (pH 2) could not remove the 3'-blocking group. Therefore the next experiments A_CE_3 and A_CE_4 were performed under already known conditions at elevated temperature (60 °C) and by using a higher amount of TBAF in THF. In case of experiment A_CE_3 we had an unexpected result as is obvious in the stacked chromatograms shown in *figure 37*: During the first 3 min of the cleavage experiment, the protected starting material was consumed in favor of **dAMP** formation.

Exper. No.	Reagent(s)/ solvent	Equiv. [mol-%]	Temp. [°C]	Time [min] for quant. CE-cleavage
A_CE_1	1 M TBAF/THF	40	40	>8
A_CE_2	0.5 M AcOH/water	217	40	No cleavage
A_CE_3	1 M TBAF/THF	40	60	Cleavage proceeds but with side reaction
A_CE_4	1 M TBAF/THF/n-propylamine (TBAF/n-propylamine 20/1)	80	60	Cleavage complete after 3 without side reaction

Chart 4: Conditions for CEM-cleavage experiments using model compound **104**

After the first 3 min, the reaction appeared to be reversible: A peak pattern with similar retention times like the one for model compound **104** increased even more during the next 6 min and didn't disappear even after 10 min complete reaction time.

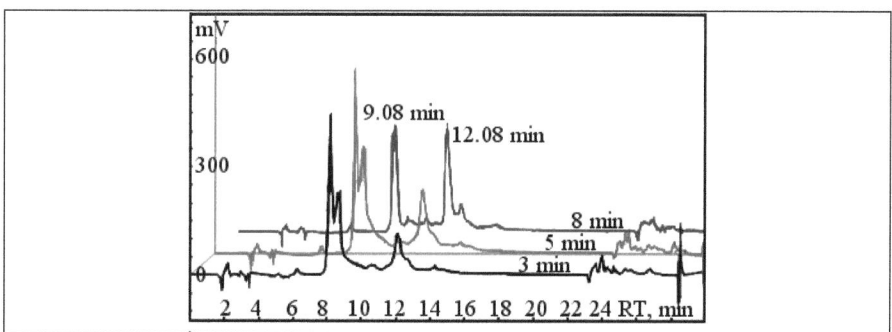

Figure 37: Stacked RP-HPLC chromatograms of cleavage experiment A_CE_3

This observation is in accord with the mechanism of the reaction which is supposed to be a reversible β-H elimination.

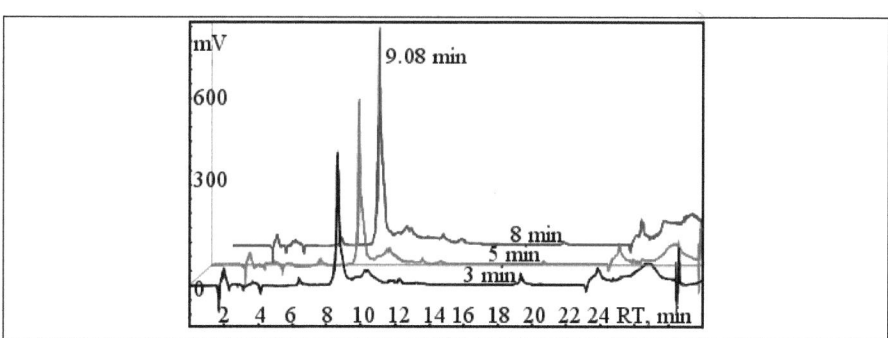

Figure 38: Stacked RP-HPLC chromatograms of cleavage experiment A_CE_4

We assume that the cleavage product acrylonitrile is still reactive by absence of excess protons but in presence of active fluoride anions. This alkene still can act as Michael acceptor and alkylate the free 3'-hydroxy function or the exocyclic amino group of the purine base. The observation of the product reversion was not made in case of CE-cleavage experiment T_CE_4 (see *chart 3*) using the model compound **95**. This nucleotide lacks any exocyclic Michael donor like the exocyclic amino function on the purine base moiety of **104** so that we assume the Michael acceptor acrylonitrile being added to the exocyclic amino group in the majority of cases.

In order to prevent any back reaction caused by reactive acrylonitrile n-propylamine was added as scavenger (see *chart 4*, entry no. A_CE_4). This trick was already applied by Saneyoshi et al. in order to enable quantitative 2'-O-CE-blocking group removal[105] in RNA-synthesis and appeared to be effective in our case, too: The starting material was deblocked after 3 min without formation of alkylated by-products or back reaction. The stacked chromatograms in *figure 38* illustrate the effective CE cleavage with absence of any peak patterns at 12 to 13 min retention but presence of 3'-deprotected compound dAMP.

5.3.4 Cyanoethyl cleavage on an oligomer

Based on the results from the monomeric model compound 3'-O-CE-dTMP **97**, we applied the optimized deblocking conditions to a short DNA oligomer (8mer) bearing the 2-cyanoethyl function on the 3'-terminus. For the evaluation of the CE cleavage on an oligodeoxynucleotide (ODN), we selected the sequence of the terminal region of the template oligomer from the polymerase acceptance tests consisting of 8 nucleotides (DNA template sequence see chapter 2.2.: Polymerase Acceptance Tests). First we synthesized the corresponding DNA 8mer without cyanoethyl function as a reference oligomer. After purification on anion exchange HPLC and desalting, the ODN could be obtained in high purity as depicted in the MALDI-mass spectrum below (see *figure 39*). The modified DNA 8mer consisting of the same sequence like the unmodified ODN was then synthesized bearing the CE function on the 3'-terminus. The resulting oligomer was purified in the same manner like the unmodified reference material giving the pure DNA 8mer ready for the cleavage experiment (see *figure 40*).

The cleavage test was then carried out in the following manner: The modified DNA 8mer was treated with a high excess (7500 equiv) of TBAF in THF/DMF 2/3 (v/v) and incubated for 15 min at 45 °C (see experimental section for details) with gentle shaking. The reaction was stopped by addition of water and the solvents were evaporated under reduced pressure (SpeedVac or freeze-drying). After desalting the crude product, OD measurement, purification on anion-HPLC and second desalting the sample was characterized via MALDI(-) TOF mass spectroscopy.

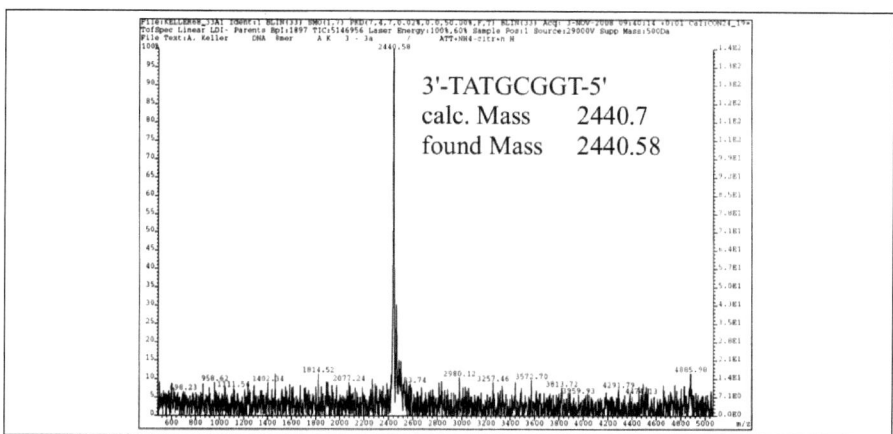

Figure 39: MALDI-MS spectrum of the unmodified DNA 8mer

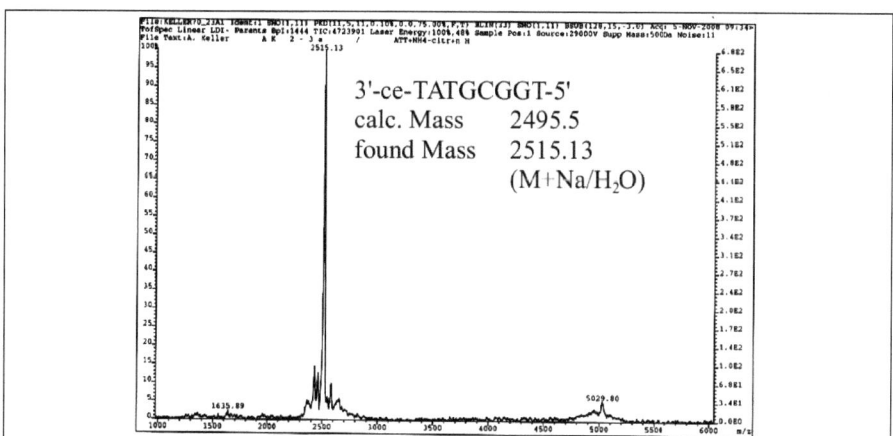

Figure 40: MALDI-MS spectrum of the modified DNA 8mer with 3'-O-CE terminus

The resulting MALDI-MS spectrum is shown in *figure 41*. The mass of the obtained peak is identical with the one of the unmodified reference DNA 8 mer which proves the efficient CE cleavage employing TBAF in THF/DMF. By comparing the result of the oligo cleavage test to the results from the monomer cleavage tests it is obvious that a significantly higher amount of cleaving agent is needed in case of the oligomer.

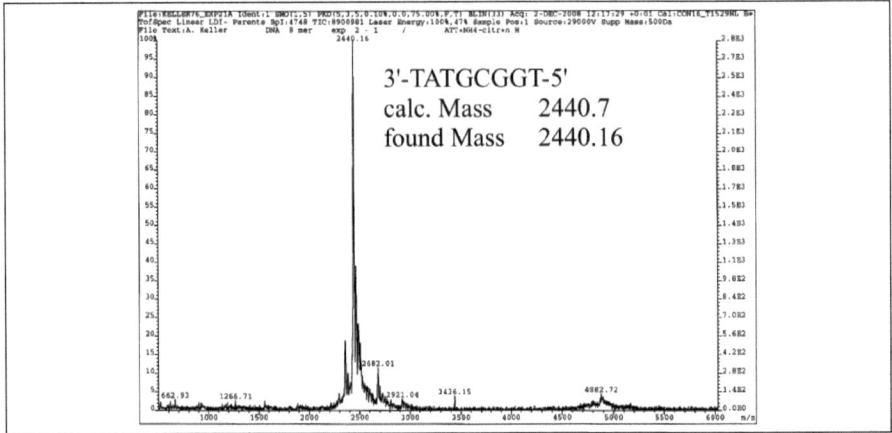

Figure 41: MALDI-MS spectrum of the modified DNA 8mer after CE cleavage

6 Summary

The polymerase acceptance tests done at Fermentas' lab using both 3'-O-CEM-dTTP and 3'-O-CE-dTTP showed that the latter nucleotide was better incorporated than the first one. Based on this knowledge all four reversible terminators containing the 2-cyanoethyl motif attached to the 3'-end were synthesized within the Array SBS project. In order to enable the attachment of a dye-linker system - an issue that is crucial for our SBS approach - the iodobases of the pyrimidines as well as of the pyrrolo[2,3-d]pyrimidines, as displayed in the first row of *figure 42*, had to be synthesized.

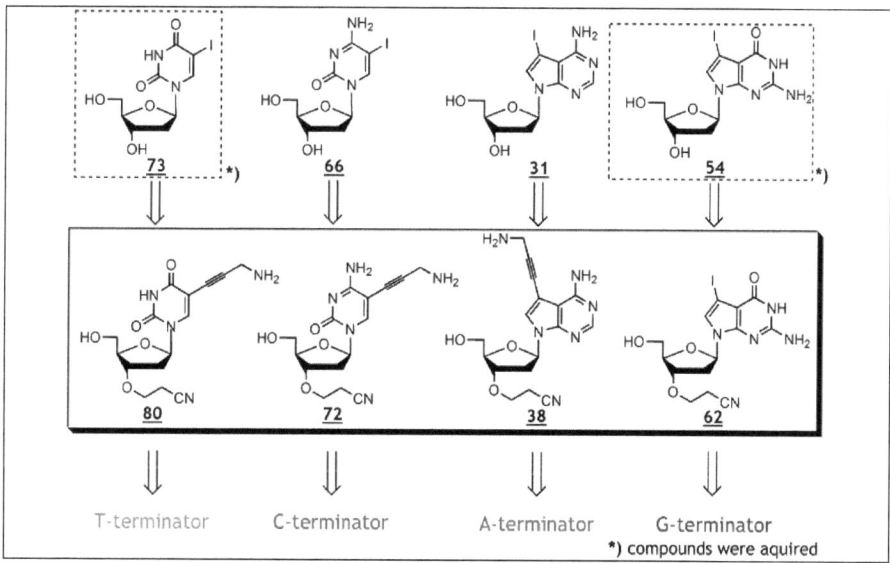

Figure 42: The iodobases and the modified key compounds needed for the synthesis of reversible terminators useful for our SBS approach

This PhD thesis reports the successful preparation of the three iodobases 4-amino-7-[2-deoxy-β-D-*erythro*-pentofuranosyl]-5-iodo-7H-pyrrolo[2,3-d]pyrimidine **31**, 5-iodo-2'-deoxycytidine **66** and 2-amino-7-[2-deoxy-β-D-*erythro*-pentofuranosyl]-5-iodo-7H-pyrrolo[2,3-d]pyrimidin-4-one **54**. For the prepara-tion of both pyrrolo[2,3-d]pyrimidine nucleobases, the glycosylation was an essential step: In both cases the best yield of the desired nucleoside was obtained when the glycosylation was carried out with sodium hydride in acetonitrile. Another issue in the preparation of these nucleosides was the synthesis of the pyrrolo[2,3-*d*]pyridine moieties that form the second building block of the nucleoside. For the synthesis of iodobase **31**, two synthetic routes were used. The first route yielded the desired heterocycle **19** (see chapters 4.3.1 and 4.3.2) in 7 % yield over five steps. The second route, which consists of six steps based on Davoll's protocol[56], turned out to be the more efficient one and gave 19 % of heterocycle **19**. After glycosylation, deprotection and reconversion of the chlorine-masked 4-amino function, nucleoside **31** was obtained in sufficient amount for further synthetic steps.

In contrast to that, the preparation of the pyrimidine base **66** was short and effective due to the fact that the non-iodinated nucleoside was commercially available. The 5-carbon-selective iodination was achieved by using Bobek's method[83b], furnishing nucleoside **66** in a good yield of 68 % over three steps.

The iodobase 5-iodo-2'-deoxyuridine **73** was commercially available from the beginning, so it was procured in sufficient amount for the elaboration of the T-terminator synthesis.

The synthesis of the iodobase 2-amino-7-[2-deoxy-β-D-*erythro*-pentofuranosyl]-5-iodo-7*H*-pyrrolo[2,3-*d*]pyrimidin-4-one **54** was the most difficult one: The first attempts to prepare this nucleobase following already published protocols[68a,68c,75,76] were not successful. Finally a complex multi-step strategy[82] with masking of both the 4-hydroxy as well as the 2-amino function of the heterocycle with subsequent glycosylation led to the desired nucleoside, but in poor yield due to big loss of material during the reconversion of the 2-amino function. In order to have enough starting material for further synthetic steps we had to acquire 2 grams of this expensive material.

With these four iodobases as starting material, the four key compounds possessing the 2-cyanoethyl group as the 3'-tag were synthesized and spectroscopically characterized. Hence this thesis reports a novel application of the 2-cyanoethyl group as 3'-O-modification for deoxynucleosides. For this purpose a smart protecting group strategy for each nucleobase enabling the 3'-selective introduction of the 2-cyanoethyl group had been designed and evaluated. The Michael addition reported by Sekine[57a] who already used this method for the preparation of 2'-O-(2-cyanoethyl) nucleotides, was the method of choice for the alkylation of the 3'-hydroxyl function. This method was applied for all four fully-protected iodo nucleobases and delivered good yields of each 3'-O-cyanoethylated product. In order to enable the attachment of a dye-linker system, the iodo function of three nucleobases was transformed into a propargyl moiety following a

Summary

standard protocol of the Sonogashira reaction. After liberation of the propargylamino function in methanolic aqueous ammonia, the three key compounds **38**, **72** and **80** were obtained in moderate yield.

The fourth key compound **62** did not need the propargylamine moiety for the next synthetic steps[55b], but still required the CE function bound to the 3'-end: Therefore we invented a novel synthetic strategy for the 3'-O-selective introduction of the CE group by using 2'-deoxyguanosine, as presented within this PhD work. The elaborated protecting group strategy was successfully transferred to the preparation of 2-amino-7-[3-O-(2-cyanoethyl)-2-deoxy-β-D-*erythro*-pento-furanosyl]-5-iodo-7H-pyrrolo[2,3-d]pyrimidin-4-one **62**. These four 3'-O-(2-cyanoethyl)-bearing key compounds (see second row in *figure 42*) were further transformed into their corresponding complete reversible terminators[55b].

Besides the synthesis of the four building blocks needed for the preparation of the complete reversible terminators, the cleavability of the CE function had to be evaluated. As mentioned previously, quantitative removal of the 3'-blocking group is essential for our SBS approach. In order to discover the conditions needed for the quantitative 3'-deblocking, three 3'-modified monophosphates as displayed in *figure 43* were designed as model compounds.

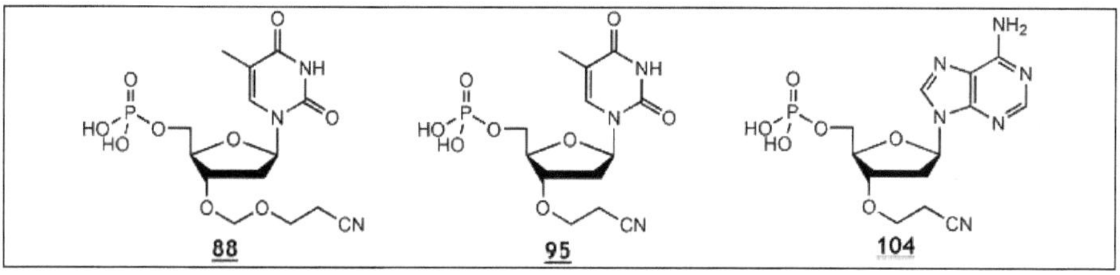

Figure 43: The three monophosphates used for kinetic cleavage tests

These previously unknown model compounds were successfully synthesized for mimicking the properties of small oligonucleotides in kinetic cleavage experiments. For each of the three monophosphates an efficient synthetic strategy has been developed. In case of the CE-bearing monophosphates **95** and **104**, the synthetic route of both compounds was quite similar to the ones used for the modification of the iodobases **31** and **73**. The synthetic strategy for the preparation of the CEM-bearing monophosphate **88** we have already published[102] and it is also reported in the diploma thesis of Angelika Keller in 2006[106]. A modified Ludwig-Eckstein triphosphate synthesis was used as effective phosphorylating method giving monophosphate **104** in moderate yield, monophosphate **88** in lower yield. For the preparation of monophosphate **95**, the conventional Yoshikawa method[92] was used. These three newly-invented nucleotides have been purified and characterized by NMR, RP-HPLC and mass spectrometry.

With these three monomers as model compounds in hand, several kinetic cleavage experiments were carried out that led to the identification of proper cleavage conditions for both 3'-modifications. In case of the CE cleavage tests, the cleavage conditions were optimized by variation of solvents, reaction temperature and the amount of cleaving agent in order to be applicable in an SBS experiment. Furthermore, we compared both the CE and the CEM group regarding their stability: Contrary to our assumption the CEM function was found to be cleavable under very similar conditions like the CE group. Quantitative cleavage of these 3'-tags was achieved when TBAF in THF was used in an excess of minimum 40 equiv, preferably at 40 °C or 60 °C. In case of the CE group, which was more interesting regarding our complete reversible terminators, the cleavage efficiency could be enhanced by addition of co-solvents like DMSO of DMF. Both 3'-blocking groups, the CE as well as the CEM group, were found to be remarkably stable under acidic conditions, but only metastable in an alkaline environment like in aqueous ammonia at elevated temperature or in aqueous potassium hydroxide. These observations suggest that the cleavage mechanism of both groups is quite similar and proceeds as β-H-elimination.

An additional CE cleavage test performed with a short oligomer revealed further results that are in accord with the results from the cleavage tests of the monomers. One big difference in the cleavage efficiency between the monomers and the oligomer is the demanded amount of cleaving agent. In case of the oligomer, 7500 equiv of TBAF in THF are needed, compared to 40 to 80 equiv for the monomer, for quantitative CE removal at 45 °C. Based on this observation it can be assumed that the solubility of the oligomer as well as the concentration of TBAF plays a very important role in the CE-cleavage efficiency. This was also demonstrated by Saneyoshi *et al.* who affirmed that the 2'-CE-cleavage from RNA dimers and oligomers is prolonged depending on the size of the oligomer[104].

However, the CE cleavage experiment employing the DNA 8mer proved that the CE function is quantitatively removable with TBAF in THF without destruction of the oligomer. Moreover, the results from the kinetic experiments carried out in strong alkaline environment confirmed the reversibility of the 2-cyanoethyl and the (2-

cyanoethoxy)methyl function as 3'-blocking groups. As a result of the extensive tests run with the monophosphate **97** we released a publication covering the monophosphate synthesis and the results of the CE cleavage experiments[107]. The CE cleavage conditions used for the DNA 8mer cleavage test are now applied to our SBS proof-of-principle using array-bound hairpin-shaped templates and are at the stage of being optimized.

6 Zusammenfassung

In Fermentas' Polymerase-Akzeptanztests, die sowohl jeweils mit 3'-O-CEM-dTTP als auch mit 3'-O-CE-dTTP als Substrat durchgeführt wurden, zeigte sich, dass letzteres der beiden Nukleotide besser eingebaut wurde als ersteres. Auf diesem Wissen basierend wurden alle vier reversiblen Terminatoren, welche die Cyanoethyl-Gruppe am 3'-Ende besitzen, im Rahmen des Array-SBS-Projekts synthetisiert. Um die – für unseren SBS Ansatz essentielle – Verknüpfung mit einem Farbstoff-Linker-System zu ermöglichen, mussten die jodierten Nukleoside der Pyrimidine als auch die der Pyrrolo[2,3-d]-pyrimidine, wie in *Abbildung 42* in der oberen Reihe gezeigt, hergestellt werden.

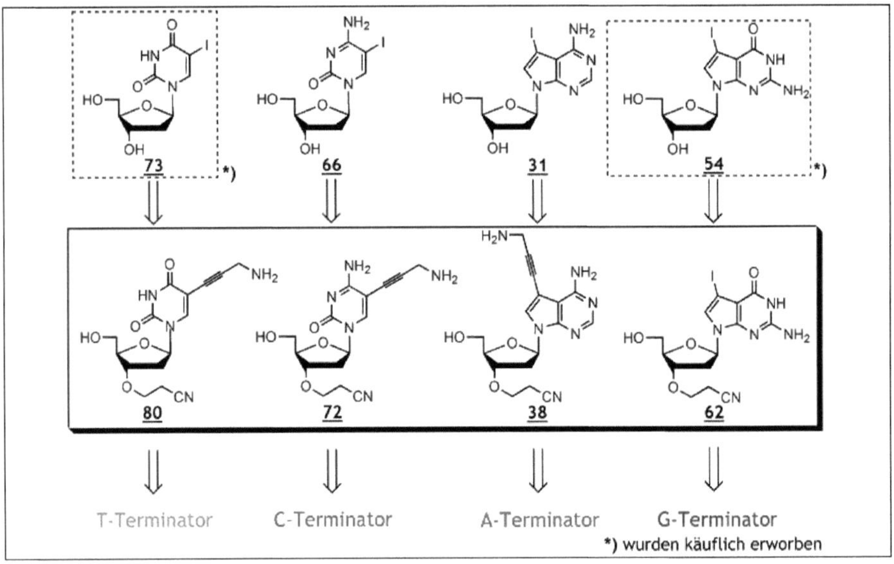

Abbildung 42: Benötigte jodierte Basen und modifizierte Schlüsselverbindungen zur Synthese reversibler Terminatoren für den Einsatz im SBS Experiment

Die vorliegende Doktorarbeit beschreibt die erfolgreiche Darstellung der drei jodierten Basen 4-Amino-7-[2-desoxy-β-D-*erythro*-pentofuranosyl]-5-jod-7*H*-pyrrolo[2,3-*d*]pyrimidin **31**, 5-Jod-2'-desoxycytidin **66** und 2-Amino-7-[2-desoxy-β-D-*erythro*-pentofuranosyl]-5-jod-7*H*-pyrrolo[2,3-*d*]pyrimidin-4-on **54**. Die Glykosylierung war dabei von zentraler

Zusammenfassung

Bedeutung in der Herstellung der beiden Pyrrolo[2,3-d]pyrimidin-Nukleoside: In beiden Fällen wurde die beste Ausbeute an gewünschtem Nukleosid erzielt, wenn die Reaktion mit Natriumhydrid in Acetonitril durchgeführt wurde. Ein weiterer wichtiger Punkt in der Darstellung dieser Nukleoside war die Synthese der Pyrrolo[2,3-*d*]pyrimidin-Einheit, die dem zweiten Bestandteil der Nukleinsäuren entspricht. Für die Synthese von Verbindung **31** wurden zwei Syntheserouten verifiziert. Syntheseroute 1 lieferte den gewünschten Heterozyklen **19** in 7 % Ausbeute über fünf Stufen. Syntheseroute 2, die über sechs Stufen verlief und auf der Veröffentlichung von Davoll[56] basierte, lieferte 19 % des Heterozyklen **19** und war damit die erfolgreichere der beiden Synthesestrategien. Nach der Glykosylierung, Entschützung und Rücktransformation der mit Chlor maskierten Aminogruppe in 4-Position wurde das Nukleosid **31** in ausreichender Menge für die nächsten Syntheseschritte erhalten.

Im Gegensatz dazu war die Darstellung der Pyrimidinbase **66** kurz und effizient aufgrund der Tatsache, dass die nichtjodierte Nucleobase käuflich zu erwerben war. Die C5-selektive Jodierung wurde unter Verwendung der Methode von Bobek[83b] erzielt, die das Nukleosid **66** in guter Ausbeute von 68 % lieferte. Die jodierte Base 5-Jod-2'-desoxyuridin war von Anfang an käuflich zu erwerben, daher stand sie uns im ausreichenden Maße zur Verfügung, um die Synthese des T-Terminators auszuarbeiten.

Die Synthese der jodierten Base 2-Amino-7-[2-desoxy-β-D-*erythro*-pentofuranosyl]-5-jod-7*H*-pyrrolo[2,3-*d*]pyrimidin-4-on war die schwierigste: Erste literaturbekannte Ansätze zur Synthese dieses Nukleosids blieben zunächst erfolglos[68a,68c,75,76]. Letztendlich führte eine komplexe mehrstufige Strategie[82], die auf der Maskierung der 4-Hydroxyl- als auch der 2-Amino-Gruppe des Heterozyklen mit anschließender Glykosylierung basiert, zum Erfolg und lieferte das erwünschte Nukleosid, jedoch in sehr niedriger Ausbeute aufgrund des hohen Substanzverlusts bei der Rücktransformation der 2-Aminogruppe. Um genügend Startmaterial für die weiteren Syntheseschritte zu besitzen, mussten 2 Gramm dieser teuren Substanz erworben werden.

Mit diesen vier jodierten Basen als Ausgangsverbindungen wurden die vier Schlüsselverbindungen, die die Cyanoethyl-Funktion als 3'-blockierende Gruppe besitzen, synthetisiert und spektroskopisch charakterisiert. Demzufolge beschreibt die vorliegende Dissertation eine neuartige Anwendung der Cyanoethyl-Gruppe als 3'-*O*-Modifikation für Desoxynukleotide. Zu diesem Zweck ist eine clevere Schutzgruppenstrategie, die die 3'-selektive Einführung der Cyanoethyl-Gruppe an dem jeweiligen Nukleosid ermöglicht, entworfen und durchgeführt worden. Die Michael Addition, welche von Sekine[57a] publiziert und zur Herstellung von 2'-O-(2-Cyanoethyl)-Nukleotiden genutzt wurde, war hierbei die Methode der Wahl zur Alkylierung der 3'-Hydroxyfunktion. Diese Methode wurde für alle komplett geschützten Jodnukleoside verwendet und lieferte das jeweilige gewünschte 3'-O-Cyanoethyl-gelabelte Produkt in guter Ausbeute. Um die Anbindung eines Fluorophor-Linker-Systems zu ermöglichen wurde die jeweilige Jodfunktion der drei Nukleobasen mittels Standardmethode der Sonogashira Reaktion in eine Propargylamin-Einheit umgewandelt. Nach Freisetzung der Propargylamin-Gruppe in wässrig methanolischer Ammoniak-lösung wurden die drei Schlüsselverbindungen **38**, **72** und **80** in mäßiger Ausbeute erhalten.

Die vierte Schlüsselverbindung **62** benötigte für die weiteren Syntheseschritte keine Propargylamin-Funktion[55b], jedoch war die Cyanoethyl-Gruppe am 3'-Ende notwendig: Hierzu entwickelten wir eine neuartige Synthesestrategie zur 3'-O-selektiven Einführung der Cyanoethyl-Gruppe unter der Verwendung von 2'-Desoxyguanosin. Die erarbeitete Schutzgruppenstrategie konnte anschließend in die Synthese zur Herstellung von 2-Amino-7-[3-O-cyanoethyl-2-desoxy-β-D-*erythro*-pentofuranosyl]-5-jod-7*H*-pyrrolo[2,3-*d*]-pyrimidin-4-on **62** übernommen werden. Diese vier 3'-O-(2-Cyanoethyl)-blockierten Schlüsselverbindungen (siehe zweite Reihe in *Abbildung 42*) wurden dann in die entsprechenden vollständigen reversiblen Terminatoren umgesetzt [55b].

Neben der Synthese der vier grundlegenden Bausteine zur Herstellung der kompletten reversiblen Terminatoren war auch die Spaltbarkeit der Cyanoethyl-Funktion zu prüfen. Wie schon vorher erwähnt wurde, ist die quantitative Abspaltung der 3'-Modifikation für unseren SBS Ansatz unabdingbar. Um die Bedingungen, die zur quantitativen Abspaltung der 3'-blockierenden Gruppe notwendig sind, herauszufinden, wurden die drei in *Abbildung 43* gezeigten 3'-modifizierten Monophosphate entworfen. Diese bislang unbekannten Modellverbindungen, welche die Eigenschaften kleiner Oligonukleotide in kinetischen Spaltungstests nachahmen sollten, wurden erfolgreich hergestellt.

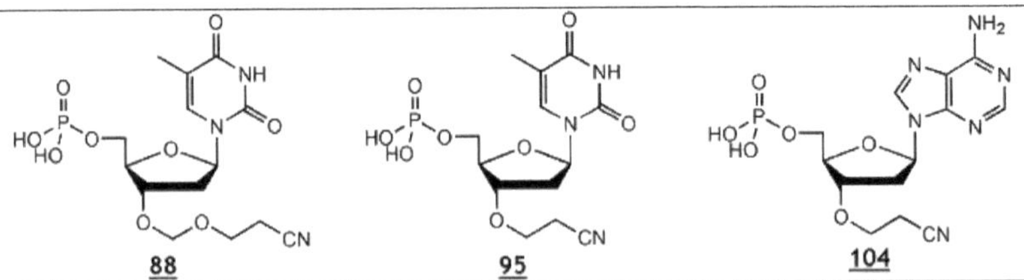

Abbildung 43: Die drei für die kinetischen Spaltungstests verwendeten Monophosphate

Zusammenfassung

Für jedes dieser drei Monophosphate wurde eine effiziente Synthesestrategie entwickelt. Im Falle der CE-gelabelten Monophosphate **95** und **104** war die jeweilige Syntheseroute den Synthesestrategien ähnlich, die schon zur Modifikation der beiden jodierten Basen **31** und **73** verwendet wurden. Wir veröffentlichen bereits die Darstellung[102] des CEM-gelabelten Monophosphats **88**, die auch in der Diplomarbeit von Angelika Keller im Jahr 2006[106] beschrieben wurde. Eine Abwandlung der Ludwig-Eckstein Triphosphatsynthese wurde als wirkungsvolle Phosphorylierungsmethode eingesetzt, die Monophosphat **104** in mäßiger und Monophosphat **88** in niedrigerer Ausbeute lieferte. Zur Herstellung des Monophosphats **95** wurde die konventionelle Methode nach Yoshikawa[92] angewandt. Diese drei neuentwickelten Nukleotide wurden aufgereinigt und mittels RP-HPLC, NMR- und Massenspektrometrie charakterisiert.

Mit den drei Monomeren in Händen wurden mehrere kinetische Spaltungsexperimente durchgeführt, die zu der Bestimmung passender Spaltungsbedingungen für beide 3'-Modifikationen führte. Im Falle der Cyanoethyl-Spaltungstests konnten die Spaltungsbedingungen für die Cyanoethyl-Gruppe sogar optimiert werden, indem das Lösungsmittel, die Reaktionstemperatur sowie die eingesetzte Menge an Spaltungsreagenz variiert wurde, um die Bedingungen für ein SBS-Experiment anwendbar zu machen.

Darüber hinaus verglichen wir die CE-Gruppe mit der CEM-Gruppe bezüglich ihrer Stabilitäten: Entgegen unserer Annahme zeigte sich, dass die CEM-Funktion unter sehr ähnlichen Bedingungen wie die CE-Gruppe gespalten wurde. Eine quantitative Spaltung dieser 3'-blockierenden Gruppen wurde durch die Verwendung von 40 bis 80 Äquivalenten als Mindestüberschuss an TBAF in THF erzielt, vorzugsweise bei 40 °C oder 60 °C. Im Falle der Cyanoethyl-Gruppe, die bezüglich der kompletten reversiblen Terminatoren die interessantere Gruppe war, konnte die Effizienz der Spaltbarkeit durch den Einsatz von Cosolventien wie DMSO oder THF noch gesteigert werden. Sowohl die CE- als auch die CEM-Gruppe waren erstaunlich stabil im sauren Milieu, allerdings nur metastabil im basischen wie z.B. in wässrig ammoniakalischer Lösung bei erhöhten Temperaturen oder in wässriger Kaliumhydroxidlösung. Diese Beobachtungen suggerieren, dass der Spaltungsmechanismus der beiden Gruppen sehr ähnlich sein und nach einer β-H-Eliminierung verlaufen muss.

Ein zusätzliches Cyanoethyl-Spaltungsexperiment, das mit einem kurzen Oligomer durchgeführt wurde, lieferte weitere Resultate, die mit den aus den Spaltungstests der Monomere gewonnenen übereinstimmen. Ein großer Unterschied zwischen den Monomeren und dem Oligomer bezüglich der Spaltungseffizienz ist die benötigte Menge an Spaltungsreagenz. Im Falle des Oligomers werden 7500 Äquivalente benötigt, um die Cyanoethyl-Gruppe bei 45 °C vollständig zu entfernen. Im Gegensatz dazu werden im Falle des Monophosphats nur 40 bis 80 Äquivalente benötigt. Basierend auf dieser Beobachtung kann man annehmen, dass sowohl die Löslichkeit des Oligomers als auch die TBAF-Konzentration eine entscheidende Rolle in der Effizienz der CE-Spaltung spielt. Dies wurde auch schon von Saneyoshi *et al.* demonstriert, der postulierte, dass im Falle der RNA die Abspaltung der Cyanoethyl-Gruppe vom 2'-Ende in Abhängigkeit von der Größe des Oligomers verzögert wird[104]. Trotzdem konnte mit dem CE-Spaltungsexperiment, welches mit dem kurzen Oligomer durchgeführt wurde, gezeigt werden, dass die CE-Gruppe mit TBAF in THF quantitativ abspaltbar ist, ohne das Oligonukleotid zu zerstören. Darüberhinaus wurde die Spaltbarkeit und damit Reversiblität der CEM- und der CE-Gruppe als 3'-blockierende Gruppen bestätigt. Als Resultat der Erkenntnisse aus den intensiven Spaltungstests, die mit Monophosphat **97** durchgeführt wurden, veröffentlichen wir eine Publikation, welche die Monophosphat-synthese und die Ergebnisse aus den CE-Spaltungsexperimenten beinhaltet[107]. Die Bedingungen aus dem CE-Oligo-Spaltungstest werden derzeit in unserem SBS-„Proof-of-Principle", welches mit Hairpin-Templaten auf einem Array durchgeführt wird, angewandt und optimiert.

7 Experimental part

7.1 Chromatography

7.1.1 Preparative Column Chromatography

Column chromatography was performed at atmospheric pressure using Silica Gel 60 from Roth.

7.1.2 Thin Layer Chromatography (TLC)

Thin layer chromatography was carried out on Silica Gel 60 F_{254}-coated aluminum sheets from Merck. The separated compounds were visualized by UV-absorption at $\lambda = 254$ nm (Camag UV-lamp).

7.1.3 Fast Protein Liquid Chromatography (FPLC)

FPLC purification of the monophosphates was performed on a Pharmacia reversed-phase FPLC system consisting of LCC-500 Plus controller, two P-500 pumps and one single-path UV Monitor UV-1. The monophosphate sample was loaded on a column packed with octadecyl-functionalized silica gel from Sigma Aldrich with peristaltic pump P1, fractions were collected with a fraction collector RediFrac. Further purification of the monophosphates was performed on RP-HPLC.

7.1.4 Buffers and methods for RP-FPLC

Preparative purification of the monophosphates was done at 4°C. The following method for the purification of the monophosphates was used:

Method FPLC-3:

Buffer A: demineralized water
Buffer B: ACN

Volume [ml]	Function	Value
0.00	CONC %B	0.0
0.00	ML/MIN	4.00
0.00	CM/ML	0.10
0.00	ML/MARK	20
0.00	CLEAR DATA	
0.00	MONITOR	1
0.00	LEVEL %	0.5
0.00	VALVE.POS	1.1
0.00	PORT.SET	6.1
0.00	PORT.SET	3.1
0.00	VALVE.POS	1.1
450.00	CONC %B	10.0
500.00	CONC %B	50.0
550.00	CONC %B	100
560.00	CONC %B	0.0
560.00	PORT.SET	3.0
560.00	PORT.SET	6.0
560.00	INTEGRATE	1
560.00	PRT PK	1.20
561.00	CONC %B	0.0
561.00	ML/MIN	0.30
561.00	HOLD	

7.1.5 Reversed-phase High Performance Liquid Chromatography (RP-HPLC)

Preparative HPLC purification of all monophosphates after synthesis was done on reversed phase using semipreparative column Phenomenex Jupiter 4µ Proteo 90 Ångstrøm (250 x 15 mm) (at 300 K in Thermotechnic-products Jetstream plus column thermostat) on a Jasco RP-HPLC system consisting of Jasco interface LC-NetII/ADC, Jasco intelligent HPLC pump PU-2080 Plus, Jasco intelligent UV/vis detector UV-2075 Plus, Jasco ternary gradient unit LG-2080-02 and Jasco 3-line degasser DG-2080-53. The crude monophosphates as well as the samples from the deprotection tests were filtered with syringe filters Spartan 13/0,45 RC (0.45 µm) from Whatman before injection. Analysis of the samples from cleavage tests was performed on analytical column Phenomenex Jupiter 4µ Proteo 90 Ångstrøm (250 x 4.6 mm) with the same Jasco HPLC system as described above. Samples were loaded by direct injection or autosampler Jasco intelligent sampler AS-950-10.

7.1.6 Buffers and methods for RP-HPLC

For analysis of the samples from cleavage experiments (loaded by auto sampler or direct injection) as well as for characterization of the monophosphates the following method was used:

Method A:
Buffer A: 1 M TEAA aqueous solution
Buffer B: demineralized water
Buffer C: ACN
Initial conditions: Flow: 1 ml/min; 10% A, 90% B, 0% C
Autosampler: Normal injection mode, Number of flush: 1
UV-Detector: wavelength 254 nm
Time table for gradient:

Time [min]	Flow [ml/min]	%A	%B	%C
20.00	1.000	10.0	60.0	30.0
21.00	1.000	0.0	0.0	100.0
25.00	1.000	0.0	0.0	100.0
26.00	1.000	10.0	90.0	0.0
30.00	1.000	10.0	90.0	0.0

For preparative purification of monophosphate **95** method B described as follows was applied:

Method B:

Buffer A: 1 M TEAA aqueous solution
Buffer B: demineralized water
Buffer C: ACN
Initial conditions: Flow: 7 ml/min; 5% A, 95% B, 0% C
UV-Detector: wavelength 254 nm
Time table for gradient:

Time [min]	Flow [ml/min]	%A	%B	%C
20.00	7.000	5.0	65.0	30.0
21.00	7.000	0.0	0.0	100.0
25.00	7.000	0.0	0.0	100.0
26.00	7.000	5.0	95.0	0.0
30.00	7.000	5.0	95.0	0.0

For preparative purification of monophosphate **97** the following method was used:

Method C:

Buffer A: 1 M TEAA aqueous solution
Buffer B: demineralized water
Buffer C: ACN
Initial conditions: Flow: 7 ml/min; 5% A, 92% B, 3% C
UV-Detector: wavelength 254 nm
Time table for gradient:

Time [min]	Flow [ml/min]	%A	%B	%C
20.00	7.000	5.0	80.0	15.0
22.00	7.000	5.0	0.0	95.0
27.00	7.000	5.0	0.0	95.0
29.00	7.000	5.0	95.0	0.0
35.00	7.000	5.0	95.0	0.0

For preparative purification of monophosphates **88** and **104** method C was used:

Method D:
Buffer A: 1 M TEAA aqueous solution
Buffer B: demineralized water
Buffer C: MeOH
Initial conditions: Flow: 6 ml/min; 5% A, 95% B, 0% C
UV-Detector: wavelength 254 nm
Time table for gradient:

Time [min]	Flow [ml/min]	%A	%B	%C
15.00	6.000	5.0	65.0	30.0
18.00	5.000	0.0	0.0	100.0
25.00	5.000	0.0	0.0	100.0
28.00	5.000	5.0	95.0	0.0
30.00	6.000	5.0	95.0	0.0

7.1.7 Anion-exchange High Performance Liquid Chromatography

The synthesized oligomers (octamers) were purified on semipreparative anion-exchange HPLC consisting of Dionex DNA Pac © PA-100 column (9 x 250 mm), Jasco intelligent UV-vis detector UV-970, Jasco intelligent HPLC pump PU-980, Jasco ternary gradient unit LG-980-02, Jasco 3-line degasser DG-980-50 and Jasco interface LC-NetII/ADC.

7.1.8 Buffers and method for anion-exchange HPLC

The oligonucleotides were purified on semipreparative anion-exchange HPLC using the following method:

Method E:
Buffer A: demineralized water
Buffer B: 0.25 M Tris-Cl aqueous solution (pH = 8)
Buffer C: 1 M NaCl aqueous solution
Initial conditions: Flow: 5 ml/min; 85% A, 10% B, 5% C
UV-Detector: wavelength 254 nm
Time table for gradient:

Time [min]	Flow [ml/min]	%A	%B	%C
20.00	5.000	65.0	10.0	25.0
23.00	5.000	0.0	0.0	100.0
25.00	5.000	0.0	0.0	100.0
27.00	5.000	85.0	10.0	5.0
30.00	5.000	85.0	10.0	5.0

7.1.9 Nuclear Magnetic Resonance (NMR) Spectroscopy

NMR spectra (^{1}H and ^{13}C) of nucleosides and other compounds were recorded in $CDCl_3$ or DMSO-d_6 on a Bruker AM 250 spectrometer operating at 250.132 MHz. NMR spectra (^{1}H, ^{13}C, ^{31}P and two-dimensional spectra) of the monophosphate, triphosphate and phosphoramidite were measured in D_2O or acetone-d_6 on a Bruker Avance 400 spectrometer operating at 400.132 MHz (161.984 MHz for coupled ^{31}P NMR, measured against 85 % phosphoric acid as external standard). Chemical shifts (δ) are given in ppm downfield from tetramethylsilane and coupling constants (J) in Hz. All spectra were measured at 300 K. Peak patterns of each signal are characterized with the following abbreviations: s for singlet, d for doublet, dd for double-doublet, t for triplet, ψt for pseudo-triplet, q for quartet and m for multiplet.

7.1.10 Mass spectrometry

Mass spectra were recorded on a Fisons MALDI VG time-of-flight (Tofspec) mass spectrometer in positive or negative mode using ATT matrix or on a Fisons Electrospray(ES) VG Platform II mass spectrometer.

7.1.11 Elementary analysis

Elementary analyses were performed on a CHN-O-Rapid analyzer from Foss-Heraeus.

7.1.12 List of chemical reagents

7.1.12.1 Chemicals for Synthesis

- Acetic acid, glacial p.a.; $C_2H_4O_2$ [60.05], bp 118 °C, d = 1.05; Merck
- Acetic acid anhydride, 99 % p.a.; $C_4H_6O_3$ [102.09], bp 139 °C, d = 1.08; Grüssing
- Acetone, p.a.; C_3H_6O [58.08], bp 56.1 °C, d = 0.791; Acros Organics
- Acetone-d_6, H_2O+D_2O < 0.02 %; euriso-top
- Acetonitrile, abs. over molecular sieves ≥99.5 %; C_2H_3N [41.05], bp 81 - 82 °C, d = 0.786; Fluka
- Acetonitrile, HPLC Gradient grade Far UV, C_2H_3N [41.05], bp 81 - 82 °C, d = 0.786; Fisher Scientific
- Acetyl chloride, 98 %; C_2H_3ClO [78.5], bp 51 °C, d = 1.1; Acros Organics
- Acrylonitrile, puriss., ≥99.5 %; C_3H_3N [53.06], bp 77°C, d = 0.806; Fluka
- 2-Amino-6-hydroxy-4-mercapto-pyrimidine monohydrate, 98 %; $C_4H_5N_3OS$ [161.18], mp > 300 °C; Sigma-Aldrich
- Ammonia, 25 % aqueous solution; H_3N [17.03]; d = 0.88; Grüssing
- Ammonia, 32 % aqueous solution; H_3N [17.03]; d = 0.88; VWR
- Ammonium acetate, ≥99.0 %; $C_2H_7NO_2$ [77.08]; mp 110 - 112 °C; Fluka
- p-Anisylchlorodiphenylmethane, 97 %; $C_{20}H_{17}ClO$ [308.80]; Acros Organics
- Argon, Quality 4.6 (99.996 Vol.-%)
- Benzoyl chloride, 99 %; C_7H_5ClO [140.57], bp 68 °C, d = 1.21; Acros Organics
- 1,8-bis-(dimethylamino)-naphthalene, ≥99.0 %; $C_{14}H_{18}N_2$ [214.31], mp 45 - 48 °C; Fluka
- N,O-Bis-(trimethylsilyl)-acetamide, purum ≥95.0 %; $C_8H_{21}NOSi_2$ [203.43], bp 71 - 73°C, d = 0.83; Fluka
- Bromoacetaldehyde diethyl acetal, techn. ≥90 %; $C_6H_{13}BrO$ [197.08]; bp 170 - 172°C, d = 1.27; Fluka
- tert-Butanol, p.a. ACS ≥99.7 %; $C_4H_{10}O$ [74.12]; bp 82 - 83°C, d = 0.79; Fluka
- 2-Butanone; C_4H_8O [72.11]; bp 80°C, d = 0.805; Fluka
- Carbon tetrachloride; CCl_4 [153.82]; bp 76 - 77 °C, d = 1.594; Merck
- Celite® 500; Fluka
- Cesium carbonate, 99.9 % (metal basis); CCs_2O_3 [325.82]; Alfa Aesar
- 3-Chloroperoxybenzoic acid, 50 - 55 % in water; $C_7H_5ClO_3$ [172.57]; mp 93 - 95 °C; Alfa

Aesar
- Chloroform-d_1, 99.80 % D; stabilized with silver foil, H_2O < 0.01 %; euriso-top
- Copper iodide, pure; CuI [190.45]; *mp* 605 °C; Riedel-de Haën
- (2-Cyanoethyl)-di(*N,N*-diisopropyl)phosphine, ≥98.0 %; $C_{15}H_{32}N_3OP$ [301.41]; *bp* 100°C, d = 0.949; Fluka
- 5-Deaza-2'-deoxy-5-iodo-guanosine; $C_{11}H_{13}IN_4O_4$ [392.15]; ChemBiotech
- 2'-Deoxyadenosine; $C_{10}H_{13}N_5O_3$ [251.24]; Pharma-Waldhof
- 2'-Deoxycytidine hydrochloride; $C_9H_{13}N_3O_4 \cdot HCl$ [263.68]; Pharma-Waldhof
- 2'-Deoxyguanosine; $C_{10}H_{13}N_5O_4$ [267.24]; Pharma-Waldhof
- 2-Deoxy-D-ribose, ≥99.0 %; $C_5H_{10}O_4$ [134.13]; *mp* 89 - 90°C; Fluka
- 2'-Deoxythymidine; $C_{10}H_{14}N_2O_5$ [242.23]; Pharma-Waldhof
- Deuterium oxide, 99.9 %; Deutero GmbH
- 2,4-Diamino-6-hydroxypyrimidine, purum ≥98 %; $C_4H_6N_4O$ [126.18]; *mp* 286 - 288°C; Fluka
- 1,2-Dichloroethane, abs. over molecular sieve ≥99.5 %; $C_2H_4Cl_2$ [98.96]; *bp* 83°C, d = 1.256; Fluka
- Dichloromethane, abs. over molecular sieve ≥99.5 %; CH_2Cl_2 [84.93]; *bp* 40°C, d = 1.325; Fluka
- 4,5-Dicyanoimidazole, purum, ≥97 %; $C_5H_2N_4$ [118.10]; *mp* 174 - 178°C; Fluka
- Dicyclohexyl-18-crown-6, purum ≥97.0 %; $C_{20}H_{36}O_6$ [372.50]; *bp* 47 - 50 °C; Fluka
- Diethyl ether, p.a.; Merck; $C_4H_{10}O$ [74.12]; *bp* 34.6 °C, d = 0.71; Merck
- Diisopropylethylamine, purum ≥98 %; $C_8H_{19}N$ [129.25]; *bp* 126 - 128°C, d = 0.755; Fluka
- *N,N*-Dimethylaniline, 99 %; $C_8H_{11}N$ [121.18]; *bp* 193°C, d = 0.95; Acros Organics
- 4-Dimethylaminopyridine, ≥99.0 %; $C_7H_{10}N_2$ [122.17]; *bp* 190 - 192 °C, d = 1.01; Fluka
- *N,N*-Dimethyl formamide, abs. over molecular sieve ≥99.8 %; C_3H_7NO [73.09]; *bp* 153°C, d = 0.944; Fluka
- *N,N*-Dimethyl formamide dimethyl acetal, purum ≥95.0 %; $C_5H_{13}NO_2$ [119.16]; *bp* 102 - 103°C, d = 0.897; Fluka
- Dimethyl sulfoxide, abs. over molecular sieve ≥99.5 %; C_2H_6OS [78.13]; *bp* 189°C, d = 1.10; Fluka
- Dimethyl sulfoxide-d_6, H_2O < 0.02 %; euriso-top
- 1,4-Dioxane, abs. over molecular sieve ≥99.5 %; $C_4H_8O_2$ [88.11]; *bp* 100 - 102°C, d = 1.034; Fluka

Experimental part

- Dowex 50W x 8 (200 - 400 mesh), p.a., H⁺-form; Fluka
- Ethanol, ≥ 99.8 %; C_2H_6O [46.07]; bp 78 °C, d = 0.79; Roth
- Ethyl cyanoacetate, purum ~ 99 %; $C_5H_7NO_2$ [113.12]; bp 207 - 210°C, d = 1.061; Fluka
- Ethylenediamine tetraacetic acid disodium-salt, analytical grade; $C_{10}H_{14}N_2O_8Na_2 \cdot 2H_2O$ [372.3]; Serva
- n-Hexane, p.a.; C_6H_{14} [86.18]; bp 69 °C, d = 0.66; Grüssing
- Hydrochloric acid, p.a. 37 %; HCl [36.46]; bp > 100 °C, d = 1.19; Merck
- 3-Hydroxypropionitrile, 98 %; C_3H_5ON [71.08]; bp 227-228°C, d = 1.041; Alfa Aesar
- Iodic acid, p.a. ≥ 99.5 %; HIO_3 [175.91]; Fluka
- Iodine; I_2 [253.81]; mp 184 °C; Fluka
- Iodomethane, 99 % stabilized; CH_3I [141.94]; bp 41 - 43°C, d = 2.28; Acros Organics
- 5-Iodo-2'-deoxyuridine; $C_9H_{11}IN_2O_5$ [354.10]; Pharma-Waldhof
- N-iodosuccinimide, ≥97 %; $C_4H_4INO_2$ [224.98]; Fluka
- Magnesium sulfate, pure 99 %; $MgSO_4$ [120.37]; Grüssing
- Methanol, abs. over molecular sieve ≥99.5 %; CH_4O [32.04]; bp 64.7°C, d = 0.791; Fluka
- Molecular sieve 3 Å; Riedel-de Haën
- Phosphorus oxychloride, ≥99 %; $POCl_3$ [153.33]; bp 105 - 110°C, d = 1.67; Riedel-de Haën
- Pivaloyl chloride, purum ≥98.0 %; C_5H_9ClO [78.13]; bp 105 - 106°C, d = 0.980; Fluka
- Potassium fluoride, 99.99 % (metal basis); KF [58.10]; mp 859°C; Alfa Aesar
- Potassium bis(trimethylsilyl)amide, 95 %; $C_6H_{18}KNSi_2$ [199.49]; Sigma-Aldrich
- Potassium hydroxide, p.a. 99 %; KOH [56.11]; mp 361 °C; Grüssing
- Potassium iodide, puriss.; KI [166.00]; mp 681 °C; Riedel-de Haën
- n-Propylamine, ≥99 %; C_3H_9N [59.11]; bp 48 °C; Sigma-Aldrich
- Pyridine, abs. over molecular sieve ≥99.8 %; C_5H_5N [79.10]; bp 115°C, d = 0.978; Fluka
- Pyridine, 99.5 %; C_5H_5N [79.10]; bp 115°C, d = 0.978; Grüssing
- syn-2-Pyridinealdoxime, 99+ %; $C_6H_6N_2O$ [122.12]; Sigma-Aldrich
- Nickel Aluminum, Raney type, Ni/Al 50/50 wt %, aqueous suspension; Acros Organics
- Sodium acetate, anhydrous, powder extra pure; $C_2H_3NaO_2$ [82.03]; mp > 300 °C; Riedel-de Haën
- Sodium carbonate, extra pure, anhydrous; Na_2CO_3 [105.99]; mp 851 °C; Merck
- Sodium chloride, ≥99.5 %; NaCl [58.44]; mp 801 °C; Fluka
- Sodium dihydrogen phosphate monohydrate, ACS; $NaH_2PO_4 \cdot H_2O$ [137.99]; mp 100 °C

(-H$_2$O); Merck
- Sodium ethoxide, 96 %, pure; C$_2$H$_5$NaO [68.04]; mp >300 °C; Acros Organics
- Sodium hydride, 57 - 63 % oil dispersion; NaH [24.00]; mp 800 °C; Lancaster
- Sodium hydrogen carbonate, pure 99 %; NaHCO$_3$ [84.01]; Grüssing
- Sodium hydroxide, 99 %, p.a.; NaOH [40.00]; mp 318 °C; Grüssing
- Sodium methoxide, purum ≥97 %; CH$_3$NaO [54.02]; mp >300 °C; Fluka
- Sodium perchlorate; NaClO$_4$ [122.44]; mp 468 °C; Sigma-Aldrich
- Sulfuryl chloride, 97 %; SO$_2$Cl$_2$ [134.97]; bp 69 °C, d = 1.68; Acros Organics
- Tetra-n-butylammonium fluoride trihydrate 98 %; C$_{16}$H$_{36}$FN·3H$_2$O [315.52]; mp 55 - 57 °C; Alfa Aesar
- Tetra-n-butylammonium fluoride 1.0 M solution in THF; C$_{16}$H$_{36}$FN [261.47]; d = 0.903; Sigma-Aldrich
- 1,1,3,3-Tetra-isopropyl-1,3-dichloro-disiloxane; Wacker
- 1,1,3,3-Tetramethylguanidine 99 %; C$_5$H$_{13}$N$_3$ [115.18]; bp 52 - 54 °C, d = 0.917; Lancaster
- Tetrakis(triphenylphosphine)palladium, ≥97 %, pure; C$_{72}$H$_{60}$P$_4$Pd [1155.56]; Fluka
- Tetrahydrofuran, abs. over molecular sieve (H$_2$O ≤ 0.005 %); C$_4$H$_8$O [72.11]; bp 65 - 67 °C, d = 0.889; Fluka
- Thiourea, p.a. ACS reagent ≥99.0 %; CN$_2$H$_4$S [76.12]; mp 174 - 177 °C; Sigma-Aldrich
- Toluene, p.a. 99.5 %; C$_7$H$_8$ [92.14]; bp 110 - 111 °C, d = 0.865; Grüssing
- p-Toluenesulfonic acid monohydrate, puriss. ≥98.5 %; C$_7$H$_8$O$_3$S·H$_2$O [190.22]; mp 102 - 105°C, Fluka
- p-Toluoyl chloride, 98 %; C$_8$H$_7$ClO [154.60]; bp 95 - 96 °C, d = 1.16; Acros Organics
- Triethylamine, p.a. ≥99.5 %; C$_6$H$_{15}$N [101.19]; bp 88 - 89°C, d = 0.727; Fluka
- Triethylamine trihydrofluoride, purum ~ 37 % HF; C$_6$H$_{15}$N·3HF [161.21]; bp 70°C; d = 0.99; Fluka
- Trifluoroacetic acid, 99 %; CF$_3$CO$_2$H [114.02]; bp 72 - 73 °C, d = 1.48; Alfa Aesar
- Trimethyl phosphate, purum ≥97 %; C$_3$H$_9$O$_4$P [140.08]; bp 192 - 194°C, d = 1.213; Fluka
- Trimethylsilyl chloride, puriss. ≥99.0 %; C$_3$H$_9$ClSi [108.64]; bp 57°C, d = 0.856; Fluka
- Tris(dioxa-3,6-heptyl)amine, 95 %, C$_{15}$H$_{33}$NO$_6$ [323.42]; bp d = 1.01; Acros Organics

7.1.12.2 Chemicals for Oligonucleotide Synthesis

Oligonucleotide synthesis was performed on a DNA-synthesizer Expedite Nucleic Acid Synthesis System from Perseptive Biosystems.
- Acetonitrile for DNA synthesis and as amidite diluents, Perseptive Biosystems
- Activator, Proligo Biochemie
- Capping Reagents (Cap A and Cap B), Proligo Biochemie
- DNA-synthesis columns (1 µmol dT 500 Å CPG) from Applied Biosystems
- 3'-phosphoramidites from Pharmacia Biotech
- 5'-phosphoramidites for the inverse synthesis (direction 5' to 3') from Glen Research
- Deblocking solution, Perseptive Biosystems
- Desalting columns PD-10 columns (prepacked with Sephadex™ G-25 M) from GE Healthcare
- Oxidizer, Biosolve

7.1.13 List of synthesized compounds

- 4-Amino-5-[3-amino-prop-1-ynyl]-7-[3-O-(2-cyanoethyl)-2-deoxy-β-D-*erythro*-pentofuranosyl)-7H-pyrrolo[2,3-d]pyrimidine **38**
- 2-Amino-7-(2-deoxy-β-D-*erythro*-pentofuranosyl)-5-iodo-7H-pyrrolo[2,3-d]pyrimidin-4-one **54**
- 6-Amino-5-(2,2-diethoxyethyl)-4-hydroxy-2-mercaptopyrimidine **22**
- 6-Amino-4-hydroxy-2-methylthiopyrimidine **15**
- 2-Amino-4-hydroxy-pyrrolo[2,3-d]pyrimidine **40**
- 4-Amino-7-[3-O-(2-cyanoethyl)-2-deoxy-β-D-*erythro*-pentofuranosyl]-5-iodo-7H-pyrrolo[2,3-d]pyrimidine **36**
- 4-Amino-7-[3-O-(2-cyanoethyl)-2-deoxy-β-D-*erythro*-pentofuranosyl]-5-(3-trifluoro-acetamido-prop-1-ynyl)-7H-pyrrolo[2,3-d]pyrimidine **37**
- 4-Amino-7-[2-deoxy-β-D-*erythro*-pentofuranosyl]-5-iodo-7H-pyrrolo[2,3-d]-pyrimidine **31**
- 5-[3-Amino-prop-1-ynyl]-3'-O-(2-cyanoethyl)-2'-deoxycytidine **72**
- 5-[3-Amino-prop-1-ynyl]-3'-O-(2-cyanoethyl)-2'-deoxyuridine **80**
- 5'-O-Benzoyl-3'-O-[(2-cyanoethoxy)methyl]-2'-deoxythymidine **85**
- 7-[5-O-Benzoyl-3-O-(2-cyanoethyl)-2-deoxy-β-D-*erythro*-pentofuranosyl]-4N-(N,N-dimethylaminomethylidenyl)-5-iodo-7H-pyrrolo[2,3-d]pyrimidine **35**
- 3N-Benzoyl-7-[3-O-(2-cyanoethyl)-2-deoxy-β-D-*erythro*-pentofuranosyl]-2N-(N,N-dimethylaminomethylidenyl)-5-iodo-7H-pyrrolo[2,3-d]pyrimidin-4-one **61**
- 3N-Benzoyl-7-[3-O-(2-cyanoethyl)-2-deoxy-5-O-(4-monomethoxytrityl)-β-D-*erythro*-pentofuranosyl]-2N-(N,N-dimethylaminomethylidenyl)-5-iodo-7H-pyrrolo[2,3-d]pyrimidin-4-one **60**
- 3N-Benzoyl-3'-O-(2-cyanoethyl)-2'-deoxy-5'-O-(4-monomethoxytrityl)thymidine **92**
- 3N-Benzoyl-3'-O-(2-cyanoethyl)-2'-deoxythymidine **93**
- 7-[5-O-Benzoyl-2-deoxy-β-D-*erythro*-pentofuranosyl]-4N-(N,N-dimethylaminomethylidenyl)-5-iodo-7H-pyrrolo[2,3-d]pyrimidine **34**
- 5'-O-Benzoyl-2'-deoxy-3'-O-(methylthiomethyl)thymidine **83**
- 3N-Benzoyl-7-[2-deoxy-5-O-(4-monomethoxytrityl)-β-D-*erythro*-pentofuranosyl]-2N-(N,N-dimethylaminomethylidenyl)-5-iodo-7H-pyrrolo[2,3-d]pyrimidin-4-one **59**
- 3N-Benzoyl-2'-deoxy-5'-O-(4-monomethoxytrityl)thymidine **91**

- 3N-Benzoyl-7-[2-deoxy-3,5-O-(tetraisopropyldisiloxane-1,3-diyl)-β-D-*erythro*-pentofuranosyl]-2N-(N,N-dimethylaminomethylidenyl)-5-iodo-7H-pyrrolo[2,3-d]pyrimidin-4-one **57**
- 3N-Benzoyl-2'-deoxythymidine **90**
- 5'-O-Benzoyl-2'-deoxythymidine **82**
- 3N-Benzoyl-2N-(N,N-dimethylaminomethylidenyl)-3'-O-(2-cyanoethyl)-2'-deoxy-5'-O-(4-monomethoxytrityl)guanosine **10a**
- 3N-Benzoyl-2N-(N,N-dimethylaminomethylidenyl)-2'-deoxyguanosine **8**
- 3N-Benzoyl-2N-(N,N-dimethylaminomethylidenyl)-2'-deoxy-5'-O-(4-monomethoxytrityl)guanosine **9a**
- 3N-Benzoyl-2N-(N,N-dimethylaminomethylidenyl)-2'-deoxy-3',5'-O-(tetraisopropyldisiloxane-1,3-diyl)guanosine **7**
- 4-Chloro-7-[2-deoxy-3,5-di-O-(4-toluoyl)-β-D-*erythro*-pentofuranosyl]-5-iodo-2-methylthio-7H-pyrrolo[2,3-d]pyrimidine **51**
- 4-Chloro-7-[2-deoxy-3,5-di-O-(4-toluoyl)-β-D-*erythro*-pentofuranosyl]-5-iodo-2-pivaloylamino-7H-pyrrolo[2,3-d]pyrimidine **46**
- 4-Chloro-7-[2-deoxy-3,5-di-O-(4-toluoyl)-β-D-*erythro*-pentofuranosyl]-5-iodo-7H-pyrrolo[2,3-d]pyrimidine **29**
- 4-Chloro-7-[2-deoxy-3,5-di-O-(4-toluoyl)-β-D-*erythro*-pentofuranosyl]-2-methylthio-7H-pyrrolo[2,3-d]pyrimidine **49**
- 4-Chloro-7-[2-deoxy-3,5-di-O-(4-toluoyl)-β-D-*erythro*-pentofuranosyl]-7H-pyrrolo[2,3-d]pyrimidine **28**
- 4-Chloro-7-[2-deoxy-β-D-*erythro*-pentofuranosyl]-5-iodo-7H-pyrrolo[2,3-d]-pyrimidine **30**
- 4-Chloro-5-iodo-2-methylthio-pyrrolo[2,3-d]pyrimidine **48**
- 4-Chloro-5-iodo-pyrrolo[2,3-d]pyrimidine **19**
- 4-Chloro-2-methylthio-pyrrolo[2,3-d]pyrimidine **47**
- 4-Chloro-pyrrolo[2,3-d]pyrimidine **18**
- 3'-O-[(2-Cyanoethoxy)methyl]-2'-deoxythymidine **86**
- 3'-O-[(2-Cyanoethoxy)methyl]-2'-deoxythymidine-5'-phosphate **88**
- 3'-O-(2-Cyanoethyl)-2'-deoxyadenosine **102**
- 3'-O-(2-Cyanoethyl)-2'-deoxyadenosine-5'-phosphate **104**
- 7-[3-O-(2-Cyanoethyl)-2-deoxy-5-O-β-D-*erythro*-pentofuranosyl]-5-iodo-7H-pyrrolo-[2,3-d]pyrimidin-4-one **62**

Experimental part

- 3'-O-(2-Cyanoethyl)-2'-deoxyguanosine **11**
- 3'-O-(2-Cyanoethyl)-2'-deoxythymidine **94**
- 3'-O-(2-Cyanoethyl)-2'-deoxythymidine-5'-(N,N-diisopropyl)phosphoramidite **96**
- 3'-O-(2-Cyanoethyl)-2'-deoxythymidine-5'-phosphate **95**
- 2'-Deoxy-(3',5'-O-diacetyl)cytidine **64**
- 7-[2-Deoxy-β-D-*erythro*-pentofuranosyl]-2N-(N,N-dimethylaminomethylidenyl)-5-iodo-7H-pyrrolo[2,3-d]pyrimidin-4-one **58**
- 2-Deoxy-3,5-di-O-p-toluoyl-α-D-*erythro*-pentofuranosyl-chloride **27**
- 7-[2-Deoxy-3,5-O-(tetraisopropyldisiloxane-1,3-diyl)-β-D-*erythro*-pentofuranosyl]-2N-(N,N-dimethylaminomethylidenyl)-5-iodo-7H-pyrrolo[2,3-d]pyrimidin-4-one **56**
- 7-[2-Deoxy-3,5-di-O-(4-toluoyl)-β-D-*erythro*-pentofuranosyl]-5-iodo-2-methylthio-7H-pyrrolo[2,3-d]pyrimidin-4-one **52**
- 7-[2-Deoxy-β-D-*erythro*-pentofuranosyl]-4N-(N,N-dimethylaminomethylidenyl)-5-iodo-7H-pyrrolo[2,3-d]pyrimidine **33**
- 2'-Deoxythymidine-5'-phosphate **97**
- 4N-(N,N-Dimethylaminomethylidenyl)-5'-O-benzoyl-3'-O-(2-cyanoethyl)-2'-deoxyadenosine **101**
- 4N-(N,N-Dimethylaminomethylidenyl)-5'-O-benzoyl-2'-deoxyadenosine **100**
- 4N-(N,N-Dimethylaminomethylidenyl)-2'-deoxyadenosine **99**
- 2N-(N,N-Dimethylaminomethylidenyl)-2'-deoxy-3,5'-O-(dibenzoyl)guanosine **9b**
- 2N-(N,N-Dimethylaminomethylidenyl)-2'-deoxyguanosine **2**
- 2N-(N,N-Dimethylaminomethylidenyl)-2'-deoxy-5'-O-(4-monomethoxytrityl)guanosine **3**
- 2N-(N,N-Dimethylaminomethylidenyl)-2'-deoxy-3',5'-O-(tetraisopropyldisiloxane-1,3-diyl)guanosine **6**
- 2,2-Dimethyl-N-(4-chloro-5-iodo-pyrrolo[2,3-d]pyrimidin-2-yl)-propionamide **45**
- 2,2-Dimethyl-N-(4-chloro-pyrrolo[2,3-d]pyrimidin-2-yl)-propionamide **44**
- 2,2-Dimethyl-N-(4-hydroxy-5-iodo-pyrrolo[2,3-d]pyrimidin-2-yl)-propionamide **42**
- 2,2-Dimethyl-N-(4-hydroxy-pyrrolo[2,3-d]pyrimidin-2-yl)-propionamide **41**
- Ethyl-(2,2-diethoxyethyl)-cyanoacetate **21**
- 4-Hydroxy-2-mercapto-pyrrolo[2,3-d]pyrimidine **23**
- 4-Hydroxy-2-methylthio-pyrrolo[2,3-d]pyrimidine **16**
- 4-Hydroxy-pyrrolo[2,3-d]pyrimidine **17**
- 5-Iodo-5'-O-benzoyl-2'-deoxyuridine **74**
- 5-Iodo-3'-O-(2-cyanoethyl)-2'-deoxycytidine **70**

- 5-Iodo-3'-O-(2-cyanoethyl)-2'-deoxy-(5'-O,3N-dibenzoyl)uridine **77**
- 5-Iodo-3'-O-(2-cyanoethyl)-2'-deoxyuridine **78**
- 5-Iodo-2'-deoxycytidine **66**
- 5-Iodo-2'-deoxy-(5'-O,3N-dibenzoyl)uridine **76**
- 5-Iodo-(3',5'-O-diacetyl)-2'-deoxycytidine **65**
- 5-Iodo-4N-*(N,N*-dimethylaminomethylidenyl)-5'-*O*-benzoyl-3'-*O*-(2-cyanoethyl)-2'-deoxycytidine **69**
- 5-Iodo-*4N-(N,N*-dimethylaminomethylidenyl)-5'-*O*-benzoyl-2'-deoxycytidine **68**
- 5-Iodo-*4N-(N,N*-dimethylaminomethylidenyl)-2'-deoxycytidine **67**
- 5-[3-Trifluoroacetamido-prop-1-ynyl]-3'-*O*-(2-cyanoethyl)-2'-deoxycytidine **71**
- 5-[3-Trifluoroacetamido-prop-1-ynyl]-3'-*O*-(2-cyanoethyl)-2'-deoxyuridine **79**

7.2 Synthesis and analytical data of all compounds

2N-(N,N-Dimethylaminomethylidenyl)-2'-deoxyguanosine

2

$C_{13}H_{18}N_6O_4$
322.32 g/mol

To a suspension of 2.67 g (10 mmol) 2'-deoxyguanosine **1** in 60 ml MeOH 5.3 ml (40 mmol) of N,N-dimethylformamide dimethyl acetal were added. The reaction mixture was agitated for 60 h at room temperature within which the crystalline product precipitated. The colorless crystals were then filtered off, washed with MeOH and dried in vacuum. The product was sufficiently pure for further synthetic use.

Yield: 2.45 g (7.6 mmol, 76 %)
TLC: R_f = 0.07 (CH_2Cl_2/MeOH: 95/5)

^{1}H-NMR: (250 MHz, DMSO-d_6)
δ [ppm] = 11.31 (br s, 1H, NH3), 8.55 (s, 1H, formamidino-CH), 8.03 (s, 1H, H6), 6.25 (dd, 1H, H1', J = 6.11 and J = 7.83), 4.38 (m, 1H, H3'), 3.83 (m, 1H, H4'), 3.54 (m, 2H, H5'), 3.16 (s, 3H, formamidino-CH$_3$), 3.03 (s, 3H, formamidino-CH$_3$), 2.64 – 2.18 (m, 2H, H2').

^{13}C-NMR: (63 MHz, DMSO-d_6)
δ [ppm] = 158.0 (C4), 157.3 (formamidino-C=N), 153.9 (C2), 149.6 (C7a), 136.6 (C6), 119.7 (C4a), 87.7 (C4'), 82.8 (C1'), 70.9 (C3'), 61.8 (C5'), 40.6 (formamidino-CH$_3$), 37.9 (C2'), 34.6 (formamidino-CH$_3$).

ESI(+)-MS: m/z 323.0 [M + H$^+$]

2N-(N,N-Dimethylaminomethylidenyl)-2'-deoxy-5'-O-(4-monomethoxytrityl)guanosine

3

$C_{33}H_{34}N_6O_5$
594.66 g/mol

2.4 g (7.45 mmol) of compound **2** was dried at 40 °C over two days in vacuum and suspended in 50 ml abs. pyridine under argon. The suspension was treated with 2.76 g (8.93 mmol) p-anisylchlorodiphenylmethane and stirred for 24 h at room temperature within which it turned into a yellow solution. The reaction was stopped by the addition of 1 ml MeOH and the mixture concentrated under reduced pressure. The yellow oily residue was taken up in 100 ml methylene chloride, washed once with 50 ml 2 % aqueous sodium carbonate solution and once with 100 ml brine. The organic layer was dried over MgSO$_4$, filtered and concentrated again. The yellow foamy product was purified via column chromatography using CH$_2$Cl$_2$/MeOH (0.5 – 5 % MeOH) as eluent.

Yield: 3.72 g (6.26 mmol, 84 %)
TLC: R$_f$ = 0.26 (CH$_2$Cl$_2$/MeOH: 95/5)

^{1}H-NMR: (400 MHz, CDCl$_3$)
δ [ppm] = 9.54 (br s, 1H, NH3), 8.50 (s, 1H, formamidino-CH), 7.70 (s, 1H, H6), 7.41 - 7.15 (m, 12H, MMT), 6.78 (ψd, 2H, MMT, J = 9.39), 6.39 (ψt, 1H, H1', J = 6.26 and J = 7.04), 4.64 (m, 1H, H3'), 4.17 (m, 1H, H4'), 3.75 (s, 3H, MMT-OCH$_3$), 3.39 – 3.26 (m, 2H, H5'), 3.02 (s, 3H, formamidino-CH$_3$), 2.99 (s, 3H, formamidino-CH$_3$), 2.55 (m, 2H, H2').

^{13}C-NMR: (100 MHz, CDCl$_3$)
δ [ppm] = 158.8 (MMT-C-OCH$_3$), 158.3 (formamidino-C=N), 158.2 (C4), 156.9 (C2), 150.4 (C7a), 144.2 (MMT), 144.1 (MMT), 136.1 (C6), 135.3 (MMT), 130.4 (MMT), 128.5 (MMT), 128 (MMT), 127.2 (MMT), 120.3 (C4a), 113.3 (MMT), 86.9 (MMT), 85.9 (C4'), 83.1 (C1'), 72.5 (C3'), 64.4 (C5'), 55.4 (MMT-OCH$_3$), 41.4 (formamidino-CH$_3$), 40.7 (C2'), 35.2 (formamidino-CH$_3$).

ESI(+)-MS: m/z 595.4 [M + H$^+$]

2N-(N,N-Dimethylaminomethylidenyl)-2'-deoxy-3',5'-O-(tetraisopropyldisiloxane-1,3-diyl)guanosine

6

$C_{25}H_{44}N_6O_5Si_2$
564.83 g/mol

802 mg (3 mmol) of 2'-deoxyguanosine **1** were coevaporated three times with dry pyridine and dried in vacuum overnight. After suspending the nucleoside in 10 ml dry pyridine, 1.06 ml (3.3 mmol) 1,1,3,3-tetraisopropyldisiloxane-1,3-dichloride were added at 0 °C. The mixture was stirred for a few minutes on cooling, then for 24 h at room temperature. After complete conversion of the starting material the reaction was stopped by addition of 10 ml MeOH. The mixture was concentrated, the residual syrup coevaporated with toluene and suspended in 20 ml methanol again. The slightly acidic reaction mixture was neutralized with triethylamine, then 2 ml (15 mmol) N,N-dimethylformamide dimethylacetal were added to the solution which was stirred further for 24 h at room temperature. After solvent removal the resulting crude product was purified on silica gel column using CH_2Cl_2/MeOH 95/5 as eluent.

Yield: 1.4 g (2.5 mmol, 83 %)
TLC: R_f = 0.49 (CH_2Cl_2/MeOH 9/1)

^{1}H-NMR: (400 MHz, $CDCl_3$)
δ [ppm] = 9.35 (br s, 1H, NH3), 8.47 (s, 1H, formamidino-CH), 7.69 (s, 1H, H6), 6.14 (dd, 1H, H1', J = 3.66 and J = 6.59), 4.58 (q, 1H, H3', J = 7.68), 3.96 - 3.84 (m, 2H, H5'), 3.75 (m, 1H, H4'), 3.07 (s, 3H, CH_3), 2.99 (s, 3H, CH_3), 2.48 - 2.44 (m, 2H, H2'), 0.98 - 0.91 (m, 28H, TIPDS-H).

^{13}C-NMR: (100 MHz, $CDCl_3$)
δ [ppm] = 158 (C4), 157.9 (formamidino-C=N), 156.8 (C2), 149.5 (C7a), 135.4 (C6), 120.3 (C4a), 85 (C4'), 81.7 (C1'), 70 (C3'), 61.8 (C5'), 41.2 (formamidino-CH_3), 40.2 (C2'), 35 (formamidino-CH_3), 17.4 - 16.8 (iso-propyl-CH_3), 13.3 - 12.4 (iso-propyl-$\underline{C}H$-CH_3).

ESI(+)-MS: m/z   565.4 [M + H$^+$]

3N-Benzoyl-2N-(N,N-dimethylaminomethylidenyl)-2'-deoxy-3',5'-O-(tetraisopropyldisiloxane-1,3-diyl)guanosine

7

$C_{32}H_{48}N_6O_6Si_2$
668.93 g/mol

1.13 g (2 mmol) of compound **6** was dissolved in 10 ml dry pyridine and cooled down to 0°C. In a syringe, 348 μl (3 mmol) of benzoyl chloride were diluted in 0.6 ml dry methylene chloride and added dropwise under argon to the nucleoside. The mixture was stirred for 30 min on cooling, then 4 h at room temperature. Another portion of 232 μl (2 mmol) benzoyl chloride in 0.6 ml methylene chloride were added at 0 °C, then the reaction was allowed to proceed overnight at room temperature. The solvent was removed, the residue taken up in 100 ml methylene chloride and washed with each 50 ml of sat. NaHCO$_3$-solution, brine and water. The organic layer was dried over MgSO$_4$, filtered and concentrated. The oily residue was purified on silica gel column (CH$_2$Cl$_2$/MeOH, 2 - 5 % MeOH) to give the pure compound as yellowish foam.

Yield: 912 mg (1.36 mmol, 68 %)
TLC: R$_f$ = 0.68 (CH$_2$Cl$_2$/MeOH: 9/1)

^{1}H-NMR: (400 MHz, CDCl$_3$)
δ [ppm] = 8.48 (s, 1H, formamidino-CH), 7.88 (ψd, 2H, benzoyl-H$_A$, J = 7.83), 7.84 (s, 1H, H6), 7.56 (ψt, 1H, benzoyl-H$_C$, J = 7.83), 7.42 (ψt, 2H, benzoyl-H$_B$, J = 7.83), 6.28 (m, 1H, H1'), 4.72 (m, 1H, H3'), 4.03 (m, 2H, H5'), 3.89 (m, 1H, H4'), 3.08 (s, 3H, formamidino-CH$_3$), 2.69 (s, 3H, formamidino-CH$_3$), 2.58 (m, 2H, H2'), 1.07 - 0.84 (m, 28H, TIPDS- H).

^{13}C-NMR: (100 MHz, CDCl$_3$)

Experimental part
132

δ [ppm] = 171.1 (benzoyl-C=O), 157.4 (C4), 156.6 (formamidino-C=N), 155.5 (C2), 148.1 (C7a), 134.9 (C6), 134.0 (benzoyl-C_C), 132.9 (benzoyl-C_{quart}), 130.3 (benzoyl-C_A), 128.9 (benzoyl-C_B), 120.4 (C4a), 85.6 (C4'), 82.8 (C1'), 70.4 (C3'), 62.1 (C5'), 41.2 (formamidino-CH_3), 39 (C2'), 35.1 (formamidino-CH_3), 17.3 – 16.8 (iso-propyl-CH_3), 13.6 – 12.7 (iso-propyl-$\underline{C}H$-CH_3).

ESI(+)-MS: m/z   669.4 [M + H$^+$]

3N-Benzoyl-2N-(*N,N*-dimethylaminomethylidenyl)-2'-deoxyguanosine

8
$C_{20}H_{22}N_6O_5$
426.43 g/mol

Under argon, 850 mg (1.27 mmol) of the nucleoside **7** were dissolved in 10 ml dry THF. Subsequently 723 µl of triethylamine tris-hydrofluoride were added in one lot and the mixture was stirred for 1.5 h at room temperature. The solvent was then evaporated and the yellow oily residue purified on column with CH_2Cl_2/MeOH 9/1 as eluent giving the pure product as glassy oil.

Yield: 532 mg (1.25 mmol, 98 %)
TLC: R_f = 0.1 (CH_2Cl_2/MeOH: 9/1)

^{1}H-NMR: (400 MHz, DMSO-d_6)
δ [ppm] = 8.58 (s, 1H, formamidino-CH), 8.18 (s, 1H, H6), 7.81 (ψd, 2H, benzoyl-H_A, J = 7.32), 7.70 (ψt, 1H, benzoyl-H_C, J = 7.32), 7.55 (ψt, 2H, benzoyl-H_B, J = 7.68), 6.32 (dd, 1H, H1', J = 6.59 and J = 7.32), 5.34 (d, 1H, 3'-OH, J = 4.03), 4.96 (t, 1H, 5'-OH, J = 5.49), 4.41 (m, 1H, H3'), 3.86 (m, 1H, H4'), 3.57 (m, 2H, H5'), 3.09 (s, 3H, CH_3), 2.63 (s, 3H, CH_3), 2.67 - 2.26 (m, 2H, H2').

^{13}C-NMR: (100 MHz, DMSO-d_6)
δ [ppm] = 171.2 (benzoyl-C=O), 157 (formamidino-C=N), 156.6 (C4), 155.1 (C2), 149.4 (C7a), 137.6 (C6), 134.4 (benzoyl-C_C), 132.9 (benzoyl-C_{quart}), 129.6

(benzoyl-C_A), 129.2 (benzoyl-C_B), 128.1 (benzoyl-C), 118.7 (C4a), 87.9 (C4'), 82.9 (C1'), 70.9 (C3'), 61.7 (C5'), 40.7 (formamidino-CH_3), 39.9 (C2'), 34.4 (formamidino-CH_3).

ESI(-)-MS: m/z 425.0 [M - H$^+$]

3N-Benzoyl-2N-(N,N-dimethylaminomethylidenyl)-2'-deoxy-5'-O-(4-monomethoxytrityl)guanosine

9a
$C_{40}H_{38}N_6O_6$
698.77 g/mol

1.06 g (2.49 mmol) of starting material **8** were dissolved under argon in 15 ml dry pyridine and treated with 1.0 g (3.24 mmol) p-anisylchlorodiphenylmethane and 30 mg (0.25 mmol) N,N-dimethylaminopyridine. The yellow solution was stirred at room temperature and quenched after 24 h reaction time by addition of 3 ml MeOH. The solvent was removed and the oily residue coevaporated three times with toluene and purified via column chromatography using CH_2Cl_2/MeOH (2 - 5 % MeOH) as eluent.

Yield: 909 mg (1.3 mmol, 52 %)
TLC: R_f = 0.25 (CH_2Cl_2/MeOH: 95/5)

^{1}H-NMR: (400 MHz, $CDCl_3$)
δ [ppm] = 8.49 (br s, 1H, formamidino-CH), 7.72 (br s, 1H, H6), 7.86 - 6.8 (m, 19H, MMT- and benzoyl-H), 6.4 (ψt, 1H, H1'), 4.65 (m, 1H, H3'), 4.18 (m, 1H, H4'), 3.77 (s, 3H, -OCH_3), 3.33 (m, 2H, H5'), 2.99 (s, 3H, formamidino-CH_3), 2.67 (s, 3H, formamidino-CH_3), 2.54 (m, 2H, H2').

^{13}C-NMR: (100 MHz, $CDCl_3$)
δ [ppm] = 170.9 (benzoyl-C=O), 158.8 (MMT-$\underline{C}$-OMe), 157.5 (C4), 156.9 (formamidino-C=N), 155.5 (C2), 148.8 (C7a), 144.3 (MMT), 144.2 (MMT), 135.3 (C6), 134.1 (benzoyl-C_C), 130.4 (MMT), 130.2 (benzoyl-C_A), 128.9 (benzoyl-C_B),

128.5 (MMT), 128 (MMT), 127.2 (MMT), 118.3 (C4a), 113.3 (MMT), 86.9 (MMT), 86.1 (C4'), 83.2 (C1'), 72.5 (C3'), 64.5 (C5'), 55.4 (MMT-O$\underline{C}$H$_3$), 41.2 (C2'), 35 (formamidino-CH$_3$), 31 (formamidino-CH$_3$).

ESI(+)-MS: m/z   699.4 [M + H$^+$]

2N-(N,N-Dimethylaminomethylidenyl)-2'-deoxy-3,5'-O-(dibenzoyl)guanosine

9b
C$_{27}$H$_{26}$N$_6$O$_6$
530.53 g/mol

Under argon, 2.0 g (4.69 mmol) of compound **8** were dissolved in 25 ml dry pyridine and cooled down to -20°C. Freshly distilled benzoyl chloride (606 μl, 5.16 mmol) was taken up in 1.5 ml dry methylene chloride and injected dropwise to the nucleoside via syringe through a septum. The reaction was allowed to proceed for 3.5 h at -20°C and stopped by the addition of 3 ml MeOH. After solvent removal, the residue was coevaporated with toluene and purified on silica gel column using CH$_2$Cl$_2$/MeOH 9/1.

Yield: 2.13 g (4.02 mmol, 86 %)
TLC: R$_f$ = 0.27 (CH$_2$Cl$_2$/MeOH: 9/1)

^{1}H-NMR: (400 MHz, DMSO-d$_6$)
δ [ppm] = 8.62 (s, 1H, formamidino-CH), 8.1 (s, 1H, H6), 8.01 - 7.36 (m, 10H, benzoyl), 6.38 (ψt, 1H, H1', J = 6.95), 4.57 (m, 1H, H4'), 4.45 (m, 2H, H5'), 4.35 (m, 1H, H3'), 3.06 (s, 3H, formamidino-CH$_3$), 2.62 (s, 3H, formamdidino-CH$_3$), 2.88 - 2.41 (m, 2H, H2').

^{13}C-NMR: (100 MHz, DMSO-d$_6$)
δ [ppm] = 171.2 (benzoyl-C=O), 165.6 (benzoyl-C=O), 157 (C4), 156.7 (formamidino-C=N), 154.9 (C2), 149.3 (C7a), 138.1 (C6), 134.4 (benzoyl-C$_{C'}$), 133.5 (benzoyl-C$_C$), 132.9 (benzoyl-C$_{quart}$), 131.6 (benzoyl), 129.6 (benzoyl-C$_{A'}$), 129.4 (benzoyl-C$_A$), 129.2 (benzoyl-C$_{B'}$), 128.8 (benzoyl-C$_B$), 118.5 (C4a), 84.3 (C4'), 83 (C1'), 70.7 (C3'), 64.4 (C5'), 40.7 (formamidino-CH$_3$), 34.4 (C2').

ESI(+)-MS: m/z   531.5 [M + H$^+$]

3N-Benzoyl-2N-(N,N-dimethylaminomethylidenyl)-3'-O-(2-cyanoethyl)-2'-deoxy-5'-O-(4-monomethoxytrityl)guanosine

10a

$C_{43}H_{41}N_7O_6$
751.83 g/mol

In an Erlenmeyer flask with triangle stirrer bar, 250 mg (0.36 mmol) of the nucleoside **9a** were dissolved under argon in 4 ml *tert*-butanol and 471 µl (7.2 mmol) freshly distilled acrylonitrile. After a few minutes, 117 mg (0.36 mmol) of cesium carbonate were added in one lot and the suspension was vigorously agitated at room temperature for 3.5 h. After consumption of the starting material, the suspension was taken up 100 ml methylene chloride and filtered over Celite. The filtrate was concentrated, the residue purified on a short silica gel column using CH_2Cl_2/MeOH 95/5 as eluent furnishing the pure product.

Yield: 218 mg (0.29 mmol, 81 %)
TLC: R_f = 0.25 (CH_2Cl_2/MeOH: 95/5)

^{1}H-NMR: (400 MHz, $CDCl_3$)
δ [ppm] = 8.51 (s, formamidino-CH), 7.87 (d, 2H, benzoyl-H$_A$, J = 7.33), 7.73 (s, 1H, H6), 7.87 (t, 1H, benzoyl-H$_C$, J = 7.33), 7.42 - 6.83 (m, 16H, trityl- and benzoyl-H), 6.35 (dd, 1H, H1', J = 5.56 and J = 8.59), 4.23 (m, 2H, H3', H4'), 3.79 (s, 3H, -OCH$_3$), 3.69 (t, 2H, -O-C$\underline{H}_2$, J = 6.06), 3.38 (m, 2H, H5'), 3.04 (s, 3H, formamidino-CH$_3$), 2.69 (s, 3H, formamidino-CH$_3$), 2.62 (t, 2H, -C$\underline{H}_2$CN, J = 6.06), 2.54 (m, 2H, H2').

^{13}C-NMR: (100 MHz, $CDCl_3$)
δ [ppm] = 170.8 (benzoyl-C=O), 158.7 (MMT-$\underline{C}$-OMe), 157.5 (C4), 156.9 (formamidino-C=N), 155.5 (C2), 148.9 (C7a), 144.3 (MMT), 144.2 (MMT), 135.3 (C6), 134.1 (benzoyl-C$_C$), 130.5 (MMT), 130.2 (benzoyl-C$_A$), 128.8 (benzoyl-C$_B$),

128.5 (MMT), 127.6 (MMT), 127.2 (MMT), 119.3 (CN), 113.3 (MMT), 86.7 (MMT), 83.5 (C4'), 82.7 (C1'), 80.6 (C3'), 63.7 (-O-$\underline{C}$H$_2$), 63.5 (C5'), 55.0 (MMT-O$\underline{C}$H$_3$), 40.4 (formamidino-CH$_3$), 40.9 (C2'), 34.6 (formamidino-CH$_3$), 18.8 (-$\underline{C}$H$_2$CN).

ESI(+)-MS: m/z 752.6 [M + H$^+$]

Elemental analysis: calculated: C, 68.69; H, 5.50; N, 13.04; O, 12.77
Found: C, 68.85; H, 5.76; N, 12.81

3'-O-(2-Cyanoethyl)-2'-deoxyguanosine

11

C$_{13}$H$_{16}$N$_6$O$_4$
320.30 g/mol

In a well-dried sealable Erlenmeyer flask with magnetic triangle stirrer bar, 1.6 g (3.02 mmol) of compound **9b** were dissolved in 4 ml (60.4 mmol) freshly distilled acrylonitrile and 8 ml *tert*-butanol under inert atmosphere. 984 mg (3.02 mmol) of cesium carbonate were added in one lot and the pale yellow suspension was vigorously stirred for about 3 h at room temperature. The mixture was diluted with 100 ml methylene chloride then, filtered over Celite, concentrated and purified on a silica gel column using CH$_2$Cl$_2$/MeOH (2 - 10 % MeOH). The resulting 700 mg (1.2 mmol) of crude compound **10b** were dissolved in a mixture of 30 ml MeOH, 10 ml methylene chloride and 20 ml 32 % aq. ammonia. The solution was stirred for 24 h at room temperature while monitoring the reaction on TLC, then the solvent was evaporated. The residue was purified on column chromatography using CH$_2$Cl$_2$/MeOH 9/1 as eluent to give product **11** as colorless crystals.

Yield: 230 mg (0.72 mmol, 24 % over two steps)
TLC: R$_f$ = 0.02 (CH$_2$Cl$_2$/MeOH: 9/1)

^{1}H-NMR: (400 MHz, DMSO-d_6)

δ [ppm] = 10.82 (br s, 1H, NH3), 7.83 (s, 1H, H6), 6.58 (br s, 2H, NH$_2$), 6.15 (dd, 1H, H1', J = 8.05, J = 2.56), 4.24 (m, 1H, H4'), 4.18 (m, 1H, H3'), 3.67 (t, 2H, -O-C$\underline{H}_2$, J = 5.86), 3.45 (m, 2H, H5'), 2.79 (t, 2H, C$\underline{H}_2$CN, J = 5.86), 2.74 – 2.32 (m, 2H, H2').

^{13}C-NMR: (100 MHz, DMSO-d_6)
δ [ppm] = 156.8 (C4), 153.7 (C2), 150.8 (C7a), 135.2 (C6), 119.1 (CN), 116.3 (C4a), 85.5 (C4'), 82.7 (C1'), 79.8 (C3'), 63.9 (-O-$\underline{C}$H$_2$), 61.6 (C5'), 37.4 (C2'), 18.2 ($\underline{C}$H$_2$CN).

ESI(+)-MS: m/z 321.0 [M + H$^+$]

6-Amino-4-hydroxy-2-methylthiopyrimidine

15
C$_5$H$_7$N$_3$OS
157.19 g/mol

16.4 g (100 mmol) of 6-amino-4-hydroxy-2-mercaptopyrimidine monohydrate **14** and 4.4 g (110 mmol) sodium hydroxide were dissolved in 200 ml water to form a clear pale yellow solution. The mixture was treated with 6.8 ml (110 mmol) of methyl iodide and stirred for 4 h at room temperature. The product precipitated as a thick cristalline mass which was filtered off, washed with 50 ml cold water and dried in vacuum for 2 days at 60 °C.

Yield: 13.3 g (84.6 mmol, 85 %)
TLC: R$_f$ = 0.22 (CH$_2$Cl$_2$/MeOH: 9/1)

^{1}H-NMR: (400 MHz, DMSO-d_6)
δ [ppm] = 11.50 (br s, 1H, H3), 6.44 (br s, 2H, NH$_2$), 4.9 (s, 1H, H5), 2.41 (s, 3H, SCH$_3$).

^{13}C-NMR: (100 MHz, DMSO-d_6)
δ [ppm] = 164.3 (C4), 163.6 (C6), 162.8 (C2), 81.2 (C5), 12.6 (CH$_3$).

ESI(+)-MS: m/z 157.6

4-Hydroxy-2-methylthio-pyrrolo[2,3-d]pyrimidine

16

$C_7H_7N_3OS$
181.21 g/mol

A mixture of 2.9 ml (18 mmol) bromoacetaldehyde diethylacetal **20**, 0.47 ml 37 % hydrochloric acid and 9.3 ml water was heated up to 90 °C for 30 min, then the clear pale yellow solution was allowed to cool down to room temperature and treated with 1.74 g (21 mmol) sodium acetate. In another flask (250 ml) a suspension of 2.5 g (16 mmol) of compound **15** and 1.08 g (13 mmol) sodium acetate in 100 ml water was heated up to 70 - 85 °C. After having reached this temperature range, the mixture from the first flask was added in one portion with vigorous stirring and heating at 80 °C continued for further 3.5 h. Progress of the reaction was obvious as the suspension turned into a solution first and, after roughly 1 h, product precipitation occurred. The crude mixture was cooled down to 0 °C for 24 h for completion of the product precipitation, then the dark yellow crystals were filtered off and washed with 100 ml water and 100 ml cold acetone. The crude product was dried in vacuum overnight and purified further on silica gel column using CH_2Cl_2/MeOH 9/1 as eluent.

Yield: 1.34 g (7.4 mmol, 46 %)
TLC: R_f = 0.46 (CH_2Cl_2/MeOH: 9/1)

^{1}H-NMR: (400 MHz, DMSO-d_6)
δ [ppm] = 12.03 (br s, 1H, H3), 11.76 (br s, 1H, H7), 6.91 (dd, 1H, H6, J = 3.28 and J = 2.53), 6.36 (dd, 1H, H5, J = 3.28 and J = 2.02), 2.52 (s, 3H, CH_3).

^{13}C-NMR: (100 MHz, DMSO-d_6)
δ [ppm] = 158.9 (C4), 154.2 (C2), 148.3 (C7a), 119.3 (C6), 104.2 (C4a), 102 (C5), 12.8 (SMe).

ESI(+)-MS: m/z 181.7

4-Hydroxy-pyrrolo[2,3-d]pyrimidine

17
$C_6H_5N_3O$
135.12 g/mol

Method A:

5.0 g (30 mmol) of compound **23** were suspended in 200 ml water and 15 ml of 28 % aqueous ammonia. 15.0 g Raney nickel (aqueous suspension) were added and the black mixture was allowed to react under reflux for 2.5 h. The dark solution was filtered while hot through Celite and the filter cake was washed with 100 ml hot water. The clear filtrate was concentrated then until pale pink crystalline product precipitated. These crystals were filtered off, washed with water and dried in vacuum at 60 °C over 2 days. Yield: 3.92 g (29 mmol, 97 %)

Method B:

5.43 g (30 mmol) of compound **16** were suspended in 250 ml EtOH and treated with 30.0 g Raney nickel. The black suspension was heated up to reflux for 3 h and filtered quickly while hot through Celite for separation from the nickel. After washing the filter cake with hot water, the slightly cloudy filtrate was concentrated until the product precipitated. The crystals were filtered off, washed with cold water and dried in vacuum at 60 °C over 2 days.

Yield: 1.49 g (11 mmol, 37 %)
TLC: R_f = 0.32 (CH$_2$Cl$_2$/MeOH: 9/1)

^{1}H-NMR: (250 MHz, DMSO-d_6)
  δ [ppm] = 11.86 (br s, 1H, H3), 11.43 (br s, 1H, H7), 7.82 (s, 1H, H2), 7.02 (d, 1H, H6, J = 3.48), 6.43 (d, 1H, H5, J = 3.48).

^{13}C-NMR: (63 MHz, DMSO-d_6)
  δ [ppm] = 158.5 (C4), 149.5 (C2), 148.1 (C7a), 120.3 (C6), 102.0 (C4a), 85.3 (C5).

ESI(+)-MS: m/z   135.7

4-Chloro-pyrrolo[2,3-d]pyrimidine

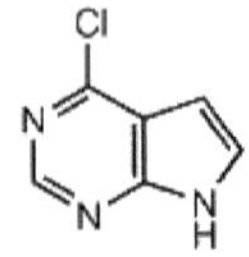

18
$C_6H_4ClN_3$
153.57 g/mol

Method A:

1.0 g (7.3 mmol) of well dried heterocycle **17** was suspended in 15 ml (160 mmol) phosphorus oxychloride and heated up to reflux for 45 min. The excess $POCl_3$ was distilled off and the reaction mixture poured on ice with strong stirring. The pH value was adjusted to 4 by addition of 10 % aqueous ammonia, the aqueous layer extracted five times with 50 ml CH_2Cl_2 each and the combined organic layers were dried over $MgSO_4$. After filtration, solvent removal and recrystallization in ethyl acetate the pure product was obtained.

Method B:

Under argon, 1.0 g (7.3 mmol) of dry heterocycle **17** was suspended in 14 ml (146 mmol) phosphorus oxychloride and 1.4 ml (11 mmol) *N,N*-dimethylaniline. The reaction mixture was heated up to reflux for 4.5 h, then the excess $POCl_3$ was distilled off and the same work-up procedure as described for method A was applied for obtaining the product as colorless crystals.

Yield: Method A: 568 mg (3.7 mmol, 51 %)
         Method B: 563 mg (3.6 mmol, 50 %)

TLC: R_f = 0.61 (CH_2Cl_2/MeOH: 9/1)

^{1}H-NMR: (250 MHz, $CDCl_3$)
   δ [ppm] = 10.69 (s, 1H, H7), 8.7 (s, 1H, H2), 7.42 (dd, 1H, H6, *J* = 2.21 and *J* = 3.48), 6.68 (dd, 1H, H5, *J* = 2.21 and *J* = 3.48).

^{13}C-NMR: (63 MHz, CDCl$_3$)
δ [ppm] = 152.5 (C4), 150.7 (C2), 150.5 (C7a), 129.3 (C6), 117.1 (C4a), 99.1 (C5).

ESI(+)-MS: m/z   153.5

4-Chloro-5-iodo-pyrrolo[2,3-d]pyrimidine

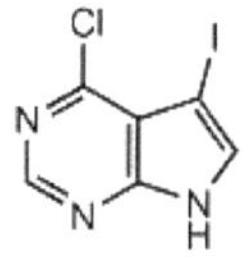

19

C$_6$H$_3$ClIN$_3$
279.47 g/mol

1.0 g (6.4 mmol) of dry compound **18** was dissolved in 100 ml dry methylene chloride, followed by the addition of 1.59 g (7 mmol) N-iodosuccinimide and stirring for 1 h at room temperature under argon atmosphere. The product precipitated as a colorless solid which was filtered off then and recrystallized in ethyl acetate. Further purification was achieved by column chromatography using CH$_2$Cl$_2$/MeOH 9/1 as eluent.

Yield: 1.59 g (5.7 mmol, 89 %)
TLC: R$_f$ = 0.64 (CH$_2$Cl$_2$/MeOH: 9/1)

^{1}H-NMR: (250 MHz, DMSO-d$_6$)
δ [ppm] = 12.95 (s, 1H, H7), 8.59 (s, 1H, H2), 7.94 (d, 1H, H6).

^{13}C-NMR: (63 MHz, DMSO-d$_6$)
δ [ppm] = 151.4 (C4), 150.6 (C2), 150.4 (C7a), 133.8 (C6), 115.7 (C4a), 51.5 (C5).

MALDI(+)-MS: m/z   279.30

Ethyl-(2,2-diethoxyethyl)-cyanoacetate

$$\text{EtOOC}\overset{\overset{\text{CN}}{|}}{\diagdown}\overset{\overset{\text{OEt}}{|}}{\diagup}\text{OEt}$$

21

$C_{11}H_{19}NO_4$
229.27

1.83 g (120 mmol) of sodium iodide, 28.05 g (200 mmol) anhydrous potassium carbonate and 31.5 ml (200 mmol) bromoacetaldehyde diethylacetal **20** were suspended in 108 ml (1 mol) ethylcyano acetate in a dry 250 ml flask with magnetic stirrer, reflux cooling and heating. The orange mixture was heated up to 130 – 150 °C for 6 h within which CO_2-gas was generated (after approximately 1.5 h the generation of CO_2-gas stopped). The dark red solution was allowed to cool down to room temperature, diluted in 160 ml water and extracted with 200 ml of diethyl ether. The aqueous layer was extracted again twice with 200 ml diethyl ether each and the combined organic layers were dried over sodium sulfate. The diethyl ether was evaporated, the residual dark red solution distilled under reduced pressure to give the product as a colorless liquid.

Fraction/substance	bp at 15 mbar	Yield
Fraction 1: 1,1-diethoxyethene *(by-product from elimination)*	47 – 68 °C	2.93 g
Fraction 2: ethylcyano acetate *(reagent and solvent)*	70 – 75 °C	65.34 g
Fraction 3: α-cyano-α-(2,2-diethoxyethyl)-acetate (product)	120 – 124 °C	29.07 g

Yield: 29.07 g (127 mmol, 63 %)

^{1}H-NMR: (250 MHz, $CDCl_3$)
δ [ppm] = 4.65 (t, 1H, O-H4-O), 4.21 (q, 2H, ester-CH_2), 3.72 – 3.42 (m, 5H, -O-CH_2(acetal), H2), 2.19 (m, 2H, H3, CH_2), 1.28 (t, 3H, ester-CH_3), 1.16 (m, 6H, acetal-CH_3).

^{13}C-NMR: (63 MHz, $CDCl_3$)
δ [ppm] = 165.9 (C1), 116.3 (CN), 99.9 (-O-C4-O), 62.8 (-O-CH_2(acetal)), 62.6 (-O-CH_2(ester)), 33.6 (C3), 33.5 (C2), 15.2 (acetal-CH_3), 15.2 (acetal-CH_3), 13.9 (ester-CH_3).

ESI(-)-MS: m/z   227.9 [M – 2H$^+$]

6-Amino-5-(2,2-diethoxyethyl)-4-hydroxy-2-mercaptopyrimidine

22

$C_{10}H_{17}N_3O_3S$
259.33 g/mol

7.61 g (100 mmol) of thiourea were dissolved under argon in 50 ml abs. EtOH, treated with 7.1 g (100 mmol) sodium ethoxide and stirred at 45 °C for 5 min at room temperature. Subsequently 22.93 g (100 mmol) of compound **21**, dissolved in 100 ml abs. EtOH were added to the ethoxide/thiourea solution and the mixture was heated up to reflux for about 4 h. After solvent removal the residue was diluted in 100 ml water and washed twice with 70 ml diethyl ether. The addition of 5.8 ml (100 mmol) glacial acetic acid caused precipitation of the colorless crystalline product from the aqueous layer. Filtering and washing the product with 50 ml cold water and 100 ml diethyl ether gave the pure compound **22**.

Yield: 18.74 g (72 mmol, 72 %)
TLC: R_f = 0.19 (CH_2Cl_2/MeOH: 9/1)

^{1}H-NMR: (400 MHz, DMSO-d_6)
  δ [ppm] = 11.70 (br s, 2H, H3, H1), 6.08 (br s, 2H, NH_2), 4.49 (t, 1H, -O-C$\underline{H}$-O, J = 5.56), 3.66 – 3.33 (m, 4H, acetal-CH_2), 2.42 (d, 2H, CH_2, J = 5.56), 1.07 (t, 6H, acetal-CH_3, J = 7.07).

^{13}C-NMR: (100 MHz, DMSO-d_6)
  δ [ppm] = 172.9 (C2), 161.9 (C4), 152.1 (C6), 101.8 (-O-$\underline{C}$(acetal)-O), 85.7 (C5), 61.7 (acetal-CH_2), 28 (C5-$\underline{C}H_2$), 15.3 (acetal-CH_3).

ESI(+)-MS: m/z   259.9

4-Hydroxy-2-mercapto-pyrrolo[2,3-*d*]pyrimidine

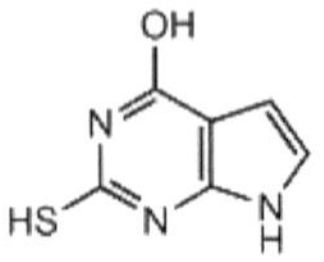

23
$C_6H_5N_3OS$
167.19 g/mol

12.96 g (50 mmol) of compound **22** were suspended in 750 ml (150 mmol) 0.2 M hydrochloric acid solution and the mixture was vigorously agitated at room temperature for 24 h. The colorless crystalline product was filtered off then, washed with 100 ml cold water and dried in vacuum at 60 °C over 2 days.

Yield: 7.96 g (47.6 mmol, 95 %)
TLC: R_f = 0.17 (CH_2Cl_2/MeOH: 9/1)

^{1}H-NMR: (400 MHz, DMSO-d_6)
δ [ppm] = 13.18 (br s, 1H, SH), 11.85 (br s, 1H, H3), 11.24 (br s, 1H, H7), 6.71 (d, 1H, H6, *J* = 3.16), 6.33 (d, 1H, H5, *J* = 3.16).

^{13}C-NMR: (100 MHz, DMSO-d_6)
δ [ppm] = 158.8 (C2), 158.7 (C4), 145.4 (C7a), 117.5 (C6), 102.2 (C5), 102 (C4a).

ESI(+)-MS: *m/z* 167.7

2-Deoxy-3,5-di-*O*-(4-toluoyl)-α-D-*erythro*-pentofuranosyl-chloride

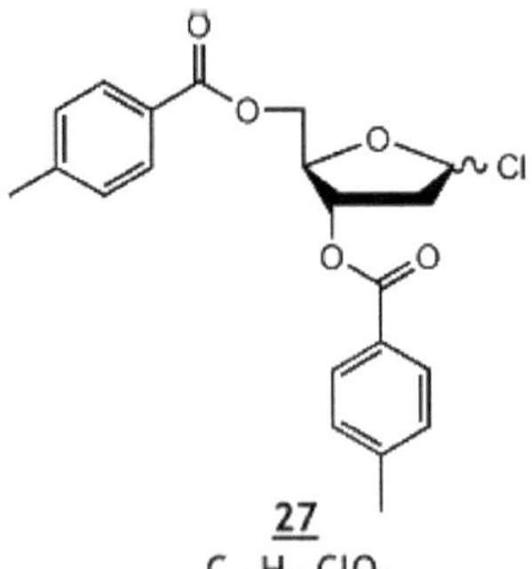

27
$C_{21}H_{21}ClO_5$
388.84 g/mol

To a solution of 10.0 g (74.5 mmol) 2-deoxy-D-ribose **24** in 120 ml abs. MeOH, 20 ml of a 1 % methanolic hydrochloric solution (prepared by adding 1.7 ml of acetyl chloride to 100 ml MeOH) were added. The reaction mixture was stirred at room temperature for 25 minutes and neutralized by addition of 4 g solid $NaHCO_3$. After filtration, the solvent was evaporated and the oily residue coevaporated three times with 50 ml dry pyridine each. The yellowish syrup was dissolved in 60 ml abs. pyridine, cooled down to 0 °C and 22 ml (160 mmol) of *p*-toluoyl chloride were added dropwise within 1 h. The reaction was allowed to proceed overnight under argon atmosphere at room temperature. The reaction mixture was diluted in 150 ml cold water then and extracted three times with 100 ml methylene chloride each. The combined organic layers were washed with 100 ml each of saturated sodium hydrocarbonate solution, 2 N hydrochloric acid and water and dried over solid sodium hydrogencarbonate. After filtration, the solvent was removed giving an oily residue. This colored syrup was dissolved in 40 ml glacial acetic acid and cooled down to 0 °C. In an Erlenmeyer flask, a saturated hydrochloric acid solution was prepared by adding 4 ml of water to well cooled 81 ml glacial acetic acid. On cooling, 16.3 ml of acetyl chloride were added in a few portions to this mixture within 5 minutes. 63 ml of this cooled saturated HCl solution were quickly filled into an ice-bath cooled flask with magnetic stirrer. The diluted sugar was poured dropwise into the saturated HCl-solution with only little stirring until a colorless crystalline precipitation occurred. The thick crystalline mass was quickly filtered off then, washed with roughly 700 ml ice cold dry diethyl ether for removing acid residues and dried in vacuum overnight. The crystalline product was stored under argon at 0 °C.

Yield: 11.68 g (30 mmol, 40 %)
TLC: R_f = 0.1 (*n*-hexane/ethyl acetate: 4/1)

^{1}H-NMR: (400 MHz, $CDCl_3$)
δ [ppm] = 8.03 (d, 2H, toluoyl-H_A, *J* = 8.05), 7.93 (d, 2H, toluoyl-H_A, *J* = 8.05), 7.26 (m, 4H, toluoyl-H_B), 6.47 (d, 1H, H1', *J* = 5.12), 5.56 (dd, 1H, H3', *J* = 2.56 and *J* = 6.95), 4.86 (q, 1H, H4', *J* = 3.29), 4.73 - 4.55 (m, 2H, H5'), 2.93 - 2.74 (m, 2H, H2'), 2.42 (s, 3H, CH_3), 2.41 (s, 3H, CH_3).

^{13}C-NMR: (100 MHz, $CDCl_3$)
δ [ppm] = 166.5 (toluoyl-C=O), 166.2 (toluoyl-C=O), 144.4 (toluoyl-C_C), 144.2 (toluoyl-C_C), 130 (toluoyl-C_A), 129.8 (toluoyl-C_A), 129.4 (toluoyl-C_B), 129.3 (toluoyl-C_B), 126.9 (toluoyl-$\underline{C}$-COO), 126.8 (toluoyl-$\underline{C}$-COO), 95.4 (C1'), 84.8 (C4'), 73.7 (C3'), 63.6 (C5'), 44.7 (C2'), 21.9 (CH_3), 21.8 (CH_3).

ESI(+)-MS: *m/z* 388.0

4-Chloro-7-[2-deoxy-3,5-di-*O*-(4-toluoyl)-β-D-*erythro*-pentofuranosyl]-7*H*-pyrrolo[2,3-*d*]pyrimidine

28
$C_{27}H_{24}ClN_3O_5$
505.95 g/mol

Under argon, 1.0 g (6.4 mmol) of the starting material **18** was dissolved in 80 ml dry acetonitrile. 0.91 g (16.2 mmol) of solid potassium hydroxide and 0.22 ml (0.64 mmol) TDA-1 were added and the mixture was stirred for 15 min at room temperature. 2.65 g (6.8 mmol) of 2-deoxy-3,5-di-*O*-(4-toluoyl)-α-D-*erythro*-pentofuranosylchloride **27** were added in one lot and the reaction mixture was stirred for 20 min at room temperature. After filtration, the solvent was evaporated and the residue purified by column chromatography (*n*-hexane/ethyl acetate = 4/1) to give the product as colorless crystals.

Yield: 1.98 g (3.9 mmol, 61 %)
TLC: R_f = 0.28 (*n*-hexane/ethyl acetate: 4/1)

^{1}H-NMR: (400 MHz, DMSO-d_6)
δ [ppm] = 8.65 (s, 1H, H2), 7.97 (d, 1H, H6, overlapped with toluoyl-H$_A$), 7.97 (d, 2H, toluoyl-H$_A$), 7.86 (d, 2H, toluoyl-H$_A$, *J* = 8.08), 7.38 (d, 2H, toluoyl-H$_B$, *J* = 8.08), 7.32 (d, 2H, toluoyl-H$_B$, *J* = 8.08), 6.77 (ψt, 1H, H1'), 6.77 (dd, 1H, H1', overlapped with H5, *J* = 6.32), 6.75 (d, 1H, H5, *J* = 3.79), 5.76 (m, 1H, H3'), 4.68 – 4.48 (m, 3H, H4', H5'), 3.18 (m, 1H, H2'), 2.76 (m, 1H, H2'), 2.41 (s, 3H, CH$_3$), 2.38 (s, 3H, CH$_3$).

^{13}C-NMR: (100 MHz, DMSO-d_6)
δ [ppm] = 165.5 (toluoyl-C=O), 165.3 (toluoyl-C=O), 151.2 (C7a), 151.0 (C4), 150.8 (C2), 144.1 (toluoyl-C$_{quart}$), 143.8 (toluoyl-C$_{quart}$), 129.5 (toluoyl-C$_A$), 129.3 (toluoyl-C$_A$), 129.3 (toluoyl-C$_B$), 129.3 (toluoyl-C$_B$), 128.8 (C6), 126.6 (toluoyl-C$_C$), 126.5 (toluoyl-C$_C$), 117.6 (C4a), 100.1 (C5), 83.9 (C1'), 81.4 (C4'),

74.9 (C3'), 64.1 (C5'), 36.0 (C2'), 21.3 (CH$_3$), 21.2 (CH$_3$).

MALDI(+)-MS: *m/z* 507.39 [M + 2H$^+$]

4-Chloro-7-[2-deoxy-3,5-di-*O*-(4-toluoyl)-β-D-*erythro*-pentofuranosyl]-5-iodo-7*H*-pyrrolo[2,3-*d*]pyrimidine

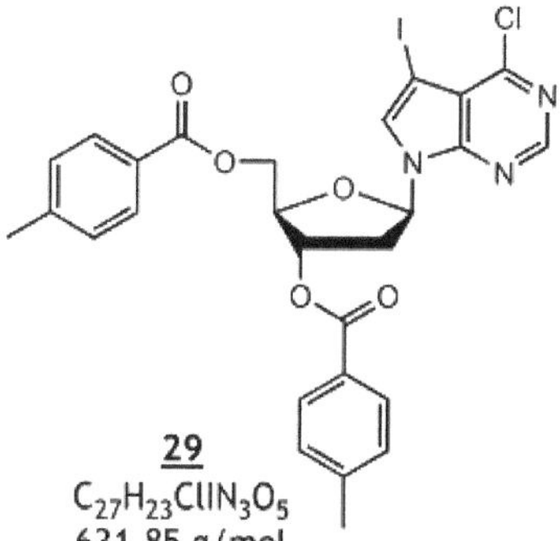

29
C$_{27}$H$_{23}$ClIN$_3$O$_5$
631.85 g/mol

Method A:

Under argon atmosphere, 980 mg (3.5 mmol) of compound **19** were dissolved in 80 ml dry acetonitrile. To this solution 0.5 g (8.8 mmol) of solid potassium hydroxide and 0.11 ml (0.35 mmol) TDA-1 were added and the mixture was stirred for 15 min at room temperature. 1.5 g (3.85 mmol) of compound **27** were added in one portion and the reaction was allowed to proceed for 20 min at room temperature. After filtration, the solvent was evaporated and the residual purified on a silica gel column (eluent: *n*-hexane/ethyl acetate = 4/1) to give the product as colorless crystals.

Yield: 663 mg (1.05 mmol, 30 %)

Method B:

430 mg (2 mmol) potassium bis(trimethylsilyl)amide were dissolved under argon in 40 ml dry ACN and treated with 44 µl (0.14 mmol) TDA-1. The mixture was stirred for 15 min at room temperature and then 381 mg (1.36 mmol) of compound **19** were added while stirring continued for another 15 min. After the addition of 580 mg (1.5 mmol) of **27** and stirring for 15 min more the product formed (as monitored by TLC). After filtration and

solvent evaporation the crude product was purified on column chromatography using n-hexane/ethyl acetate = 4/1 as eluent.

Yield: 644 mg (1.02 mmol, 51 %)

Method C:
1.0 g (3.6 mmol) of compound **19** was dissolved under argon in 50 ml dry ACN and stirred at 40 - 50 °C until complete solvation occured. Then the mixture was treated with 155 mg (3.6 mmol) sodium hydride (60 % suspension in mineral oil) and stirred for 15 min at room temperature. Subsequently, 1.36 g (3.6 mmol) of compound **27** were added and the reaction mixture was stirred for 2 h at room temperature. The reaction was stopped by the addition of Dowex (H⁺-form), then the mixture was taken up in 100 ml methylene chloride and filtered over Celite. The filtrate was concentrated and the resulting yellow foamy crude product purified on column chromatography (n-hexane/ethyl acetate 4/1) to give the product as colorless crystals.

Yield: 1.84 g (2.9 mmol, 81 %)
TLC: R_f = 0.31 (n-hexane/ethyl acetate: 4/1)

^{1}H-NMR: (400 MHz, DMSO-d_6)
δ [ppm] = 8.65 (s, 1H, H2), 8.18 (s, 1H, H6), 7.95 (d, 2H, toluoyl-H_A, J = 8.08), 7.85 (d, 2H, toluoyl-H_A, J = 8.08), 7.36 (d, 2H, toluoyl-H_A, J = 8.08), 7.31 (d, 2H, toluoyl-H_A, J = 8.08), 6.76 (dd, 1H, H1', J = 6.32, J = 7.58), 5.75 (m, 1H, H3'), 4.69 - 4.38 (m, 3H, H4', H5'), 3.09 (m, 1H, H2'), 2.77 (m, 1H, H2'), 2.40 (s, 3H, CH_3), 2.38 (s, 3H, CH_3).

^{13}C-NMR: (100 MHz, DMSO-d_6)
δ [ppm] = 165.4 (toluoyl-C=O), 165.2 (toluoyl-C=O), 151.3 (C2), 150.8 (C4), 150.6 (C7a), 144.1 (toluoyl-C_{quart}), 143.8 (toluoyl-C_{quart}), 133.3 (C6), 129.5 (toluoyl-C_A), 129.3 (toluoyl-C_A), 129.3 (toluoyl-C_B), 129.2 (toluoyl-C_B), 126.5 (toluoyl-C_C), 126.4 (toluoyl-C_C), 116.8 (C4a), 83.7 (C1'), 81.6 (C4'), 74.7 (C3'), 63.9 (C5'), 54.1 (C5), 36.2 (C2'), 21.3 (CH_3), 21.2 (CH_3).

MALDI(+)-MS: m/z 633.15 [M + 2H⁺]

4-Chloro-7-[2-deoxy-β-D-*erythro*-pentofuranosyl]-5-iodo-7*H*-pyrrolo[2,3-*d*]-pyrimidine

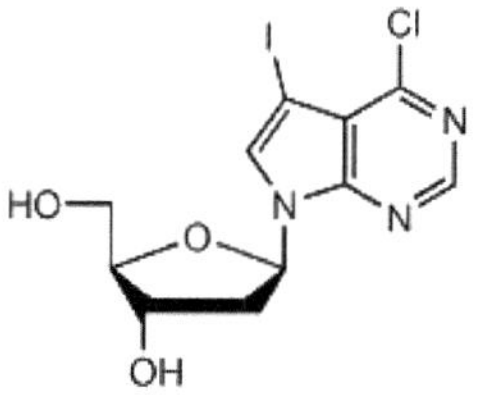

30
$C_{11}H_{11}ClIN_3O_3$
395.58 g/mol

In a sealable 100 ml flask, 1.26 g (2 mmol) of nucleoside **29** were dissolved in 30 ml sat. ammonia and 50 ml MeOH, then stirred at room temperature for 20 h. After solvent removal the residue was purified on silica gel column in CH_2Cl_2/MeOH (2 - 10 % MeOH) to give the product as colorless crystals.

Yield: 752 mg (1.9 mmol, 95 %)
TLC: R_f = 0.35 (CH_2Cl_2/MeOH: 9/1)

^{1}H-NMR: (250 MHz, DMSO-d_6)
δ [ppm] = 8.66 (s, 1H, H6), 8.21 (s, 1H, H2), 6.63 (ψt, 1H, H1'), 5.32 (d, 1H, 3'-OH, J = 4.39), 4.99 (t, 1H, 5'-OH, J = 5.49), 4.37 (m, 1H, H3'), 3.85 (m, 1H, H4'), 3.56 (m, 2H, H5'), 2.52 (m, 1H, H2'), 2.26 (m, 1H, H2').

^{13}C-NMR: (63 MHz, DMSO-d_6)
δ [ppm] = 151.1 (C2), 150.6 (C4), 150.4 (C7a), 133.4 (C6), 116.6 (C4a), 87.7 (C1'), 83.4 (C4'), 70.6 (C3'), 61.5 (C5'), 53.4 (C5), 39.8 (C2').

ESI(-)-MS: *m/z* 393.37 [M - 2H$^+$]

4-Amino-7-[2-deoxy-β-D-*erythro*-pentofuranosyl]-5-iodo-7*H*-pyrrolo[2,3-*d*]-pyrimidine

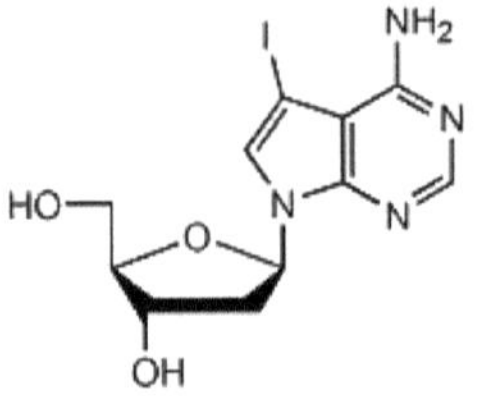

31
$C_{11}H_{13}IN_4O_3$
376.15 g/mol

Method A:

593 mg (1.5 mmol) of nucleoside **30** were suspended in 100 ml 25 % aqueous ammonia and filled into a Parr steel bomb. The mixture was stirred at 110 °C (4 bar) for about 15 h within which the white crystalline starting material completely dissolved. The yellow solution was concentrated then and the residue purified by column chromatography in CH_2Cl_2/MeOH 9/1.

Yield: 480 mg (1.28 mmol, 85 %)

Method B:

In a sealed microwave vessel, 440 mg (1.11 mmol) of compound **30** were suspended in 40 ml 25 % aqueous ammonia and stirred at 110 °C (2 bar, 150 W) for about 3 h within which the colorless crystalline starting material fully dissolved. The pale yellow solution was concentrated and the crude product purified on column chromatography in CH_2Cl_2/MeOH 9/1.

Yield: 346 mg (0.92 mmol, 83 %)
TLC: R_f = 0.23 (CH_2Cl_2/MeOH: 9/1)

^{1}H-NMR: (400 MHz, DMSO-d_6)
δ [ppm] = 8.10 (s, 1H, H2), 7.65 (s, 1H, H6), 7.35 (br s, 1H, NH$_2$(exchange)), 6.65 (br s, 1H, NH$_2$(exchange)), 6.48 (ψt, 1H, H1'), 5.26 (d, 1H, 3'-OH, J = 4.05), 5.03 (t, 1H, 5'-OH, J = 5.5), 4.33 (m, 1H, H3'), 3.82 (m, 1H, H4'), 3.54 (m, 2H, H5'), 2.54 – 2.15 (m, 2H, H2').

^{13}C-NMR: (100 MHz, DMSO-d_6)
δ [ppm] = 157.2 (C4), 151.9 (C2), 149.7 (C7a), 126.7 (C6), 103.1 (C4a), 87.4

(C4'), 82.9 (C1'), 70.9 (C3'), 61.8 (C5'), 51.7 (C5), 39.7 (C2').

ESI(-)-MS: m/z   374.9 [M - 2H$^+$]

7-[2-Deoxy-β-D-*erythro*-pentofuranosyl]-4N-(*N,N*-dimethylaminomethylidenyl)-5-iodo-7*H*-pyrrolo[2,3-*d*]pyrimidine

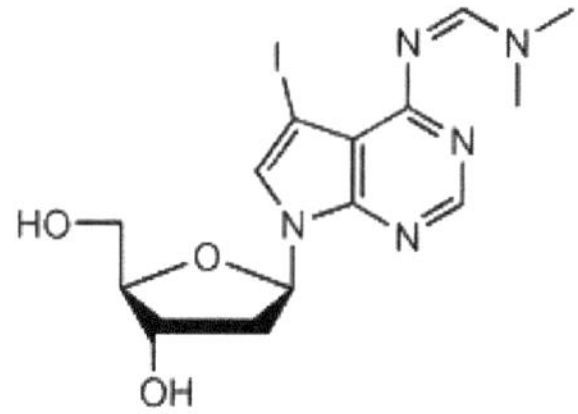

33
C$_{14}$H$_{18}$IN$_5$O$_3$
431.23 g/mol

2.07 g (5.5 mmol) of nucleoside **31** were dissolved under argon in 15 ml dry DMF and 11 ml (82.5 mmol) *N,N*-dimethylformamide dimethylacetal. The mixture was stirred at 55 °C for 2 h and concentrated. The residue was purified on silica gel column using CH$_2$Cl$_2$/MeOH (5 – 10 % MeOH) as eluent to furnish the product as yellow crystals.

Yield: 1.91 g (4.4 mmol, 80 %)
TLC: R$_f$ = 0.28 (CH$_2$Cl$_2$/MeOH: 9/1)

^{1}H-NMR: (400 MHz, DMSO-*d*$_6$)
δ [ppm] = 8.82 (s, 1H, formamidino-CH), 8.31 (s, 1H, H2), 7.71 (s, 1H, H6), 6.53 (dd, 1H, H1', *J* = 5.87 and *J* = 8.07), 4.34 (m, 1H, H3'), 3.82 (m, 1H, H4'), 3.55 (m, 2H, H5'), 3.22 (s, 3H, formamidino-CH$_3$), 3.18 (s, 3H, formamidino-CH$_3$), 2.48 – 2.16 (m, 2H, H2').

^{13}C-NMR: (100 MHz, DMSO-*d*$_6$)
δ [ppm] = 160.2 (C4), 156.2 (formamidino-C=N), 151.4 (C2), 150.8 (C7a), 128.5 (C6), 110.2 (C4a), 87.4 (C4'), 83 (C1'), 71 (C3'), 61.9 (C5'), 53.6 (C5), 40.4 (formamidino-CH$_3$), 39.9 (C2'), 34.9 (formamidino-CH$_3$).

ESI(-)-MS: m/z   430.1 [M - H$^+$]

7-[5-O-Benzoyl-2-deoxy-β-D-*erythro*-pentofuranosyl]-4N-(N,N-dimethylaminomethylidenyl)-5-iodo-7H-pyrrolo[2,3-d]pyrimidine

34
$C_{21}H_{22}IN_5O_4$
535.33 g/mol

1.72 g (4 mmol) of nucleoside **33** was coevaporated three times and dissolved in 80 ml dry pyridine. After cooling the yellow solution down to -20 °C, 488 µl (4.2 mmol) of benzoyl chloride in 20 ml dry methylene chloride were added dropwise via syringe. The mixture was stirred at -20 °C for 1 h and the reaction progress monitored via TLC. After completion of the reaction, 5 ml MeOH were added and the solvent was removed under reduced pressure. The oily residue was coevaporated with toluene and purified on column chromatography with CH_2Cl_2/MeOH (2 - 5 % MeOH) to give the product as yellowish crystals.

Yield: 1.76 g (3.29 mmol, 82 %)
TLC: R_f = 0.06 (CH_2Cl_2/MeOH: 95/5)

^{1}H-NMR: (400 MHz, DMSO-d_6)
δ [ppm] = 8.81 (s, 1H, formamidino-CH), 8.4 (s, 1H, H2), 7.98 (ψd, 2H, benzoyl-H$_A$, J = 7.09), 7.74 (s, 1H, H6), 7.67 (t, 1H, benzoyl-H$_C$, J = 7.58), 7.54 (t, 2H, benzoyl-H$_B$, J = 7.09), 6.58 (ψt, 1H, H1', J = 6.85), 4.56 - 4.38 (m, 3H, H3', H5'), 4.11 (m, 1H, H4'), 3.24 (s, 3H, formamidino-CH$_3$), 3.22 (s, 3H, formamidino-CH$_3$), 2.71 - 2.28 (m, 2H, H2').

^{13}C-NMR: (100 MHz, DMSO-d_6)
δ [ppm] = 165.6 (benzoyl-C=O), 160 (C4), 156.8 (formamidino-C=N), 151.4 (C2), 150.4 (C7a), 133.4 (benzoyl-C$_C$), 129.4 (benzoyl-C$_{quart}$), 129.2 (benzoyl-C$_A$), 128.9 (benzoyl-C$_B$), 128.2 (C6), 109.9 (C4a), 83.7 (C4'), 82.7 (C1'), 70.5 (C3'), 64.5 (C5'), 54.8 (C5), 40.6 (formamidino-CH$_3$), 39 (C2'), 35 (formamidino-CH$_3$).

ESI(+)-MS: m/z 536.1 [M + H$^+$]

7-[5-O-Benzoyl-3-O-(2-cyanoethyl)-2-deoxy-β-D-*erythro*-pentofuranosyl]-4N-(N,N-dimethylaminomethylidenyl)-5-iodo-7H-pyrrolo[2,3-d]pyrimidine

35
$C_{24}H_{25}IN_6O_4$
588.40 g/mol

In an Erlenmeyer flask with triangle stirring bar, 1.34 g (2.5 mmol) of well-dried nucleoside **34** were dissolved in 10 ml *tert*-butanol and in 3.3 ml (50 mmol) freshly distilled acrylonitrile. 814 mg (2.5 mmol) of cesium carbonate were added and the mixture was stirred vigorously under argon 3 h at room temperature. The mixture was filtered through Celite, the filter cake washed with methylene chloride and the organic layer was concentrated then. The residue was purified on a short silica gel column using CH_2Cl_2/MeOH (5 – 10 % MeOH) to give the product as yellowish foam.

Yield: 1.32 g (2.24 mmol, 90 %)
TLC: R_f = 0.21 (CH_2Cl_2/MeOH: 95/5)

^{1}H-NMR: (400 MHz, $CDCl_3$)
δ [ppm] = 8.75 (s, 1H, formamidino-CH), 8.40 (s, 1H, H2), 8.02 (ψd, 2H, benzoyl-H_A, J = 7.09), 7.58 (m, 1H, benzoyl-H_C, J = 7.34), 7.47 (t, 2H, benzoyl-H_B, J = 7.09), 7.24 (s, 1H, H6), 6.59 (ψt, 1H, H1', J = 6.6), 4.56 (m, 2H, H5'), 4.35 (m, 2H, H3',H4'), 3.72 (t, 2H, -O-C$\underline{H}_2$, J = 6.36), 3.28 (s, 3H, formamidino-CH$_3$), 3.16 (s, 3H, formamidino-CH$_3$), 2.62 (t, 2H, -C$\underline{H}_2$CN, J = 6.36), 2.57 (m, 2H, H2').

^{13}C-NMR: (100 MHz, $CDCl_3$)
δ [ppm] = 166.4 (benzoyl-C=O), 160.5 (C4), 156.1 (formamidino-C=N), 151.9 (C2), 151.2 (C7a), 133.5 (benzoyl-C_C), 129.7 (benzoyl-C_A), 129.6 (benzoyl-C_{quart}), 128.8 (benzoyl-C_B), 127.7 (C6), 117.5 (CN), 111.6 (C4a), 84 (C1'), 81.9 (C4'), 80.4 (C3'), 64.5 (C5'), 64.4 (O-CH$_2$), 53.8 (C5), 41 (formamidino-CH$_3$), 37.7 (C2'), 35.6 (formamidino-CH$_3$), 19.2 ($\underline{C}H_2$CN).

ESI(+)-MS: m/z 589.2 [M + H$^+$]

4-Amino-7-[3-O-(2-cyanoethyl)-2-deoxy-β-D-*erythro*-pentofuranosyl]-5-iodo-7*H*-pyrrolo[2,3-*d*]pyrimidine

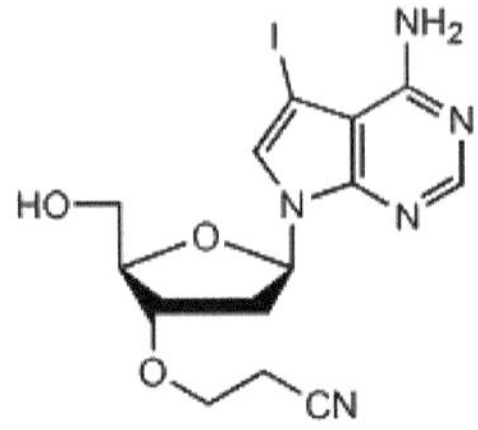

36
$C_{14}H_{16}IN_5O_3$
429.21 g/mol

1.12 g (2 mmol) of nucleoside **35** were dissolved in 70 ml sat. methanolic ammonia and 10 ml 32 % aqueous ammonia. The mixture was stirred at 50 °C and concentrated after complete conversion of the starting material (15 h). The crystalline product was obtained as colorless crystals after purification on a short silica gel column using CH_2Cl_2/MeOH (10 – 15 % MeOH).

Yield: 644 mg (1.5 mmol, 75 %)
TLC: R_f = 0.29 (CH_2Cl_2/MeOH: 9/1)

^{1}H-NMR: (400 MHz, DMSO-d_6)
δ [ppm] = 8.10 (s, 1H, H2), 7.68 (s, 1H, H6), 6.68 (br s, 2H, NH$_2$), 6.44 (dd, 1H, H1', *J* = 5.87 and *J* = 8.56), 5.14 (t, 1H, 5'-OH, *J* = 5.5), 4.23 (m, 1H, H3'), 3.96 (m, 1H, H4'), 3.67 (t, 2H, -O-C$\underline{H}_2$, *J* = 6.11), 3.55 (m, 2H, H5'), 2.81 (t, 2H, C$\underline{H}_2$CN, *J* = 6.11), 2.59 – 2.31 (m, 2H, H2').

^{13}C-NMR: (100 MHz, DMSO-d_6)
δ [ppm] = 157.2 (C4), 152 (C2), 149.8 (C7a), 126.7 (C6), 119.2 (CN), 103.1 (C4a), 84.6 (C4'), 82.9 (C1'), 79.9 (C3'), 63.4 (O-$\underline{C}$H$_2$), 61.8 (C5'), 52.1 (C5), 36.5 (C2'), 18.2 ($\underline{C}$H$_2$CN).

ESI(+)-MS: *m/z* 429.9 [M + H$^+$]

4-Amino-7-[3-O-(2-cyanoethyl)-2-deoxy-β-D-*erythro*-pentofuranosyl]-5-(3-trifluoroacetamido-prop-1-ynyl)-7H-pyrrolo[2,3-d]pyrimidine

37
$C_{19}H_{19}F_3N_6O_4$
452.39 g/mol

A solution of 350 mg (0.82 mmol) of nucleoside **36** and 567 µl (4.1 mmol) triethylamine in 13 ml dry DMF was degassed three times, then 94 mg (0.082 mmol) Pd(PPh$_3$)$_4$ and 31 mg (0.164 mmol) of copper iodide were added. The mixture was treated with 290 µl (1.63 mmol) TFA-protected propargylamine and stirred 3 h under argon at room temperature in the dark. After solvent evaporation, the dark-yellow residue was taken up in 100 ml methylene chloride and washed twice with 50 ml 5 % aq. disodium-EDTA solution. The aqueous layer was extracted again with 100 ml 2-butanone, then the combined organic layers were dried over MgSO$_4$, filtered and concentrated. The residue was purified further on a short flash column using CH$_2$Cl$_2$/MeOH (5 – 10 % MeOH) furnishing the product as yellow-brownish oil.

Yield: 257 mg (0.57 mmol, 69 %)
TLC: R$_f$ = 0.23 (CH$_2$Cl$_2$/MeOH: 9/1)

^{1}H-NMR: (250 MHz, DMSO-d$_6$)
δ [ppm] = 10.1 (br t, 1H, TFA-N$\underline{H}$, J = 5.14), 8.18 (br s, 1H, H2), 7.79 (s, 1H, H6), 6.85 (br s, 2H, NH$_2$), 6.42 (dd, 1H, H1', J = 5.87 and J = 7.83), 5.17 (t, 1H, 5'-OH, J = 5.38), 4.31 (d, 2H, propargyl-C$\underline{H}_2$, J = 5.38), 4.25 (m, 1H, H3'), 3.98 (m, 1H, H4'), 3.67 (t, 2H, -O-C$\underline{H}_2$, J = 6.11), 3.55 (m, 2H, H5'), 2.81 (t, 2H, C$\underline{H}_2$CN, J = 6.11), 2.61 – 2.31 (m, 2H, H2').

^{13}C-NMR: (63 MHz, DMSO-d$_6$)

δ [ppm] = 157.4 (C4), 156.6 (TFA-C=O), 152.6 (C2), 149.3 (C7a), 126.5 (C6), 119.2 (CN), 117.6 (CF₃), 102.2 (C4a), 94.3 (C5), 86.8 (propargyl-C-CH₂), 84.7 (C4'), 83.2 (C1'), 79.8 (C3'), 76.1 (propargyl-C-C5), 63.4 (O-CH₂), 61.8 (C5'), 36.5 (C2'), 30.7 (propargyl-CH₂), 18.2 (CH₂CN).

ESI(+)-MS: m/z   453.2 [M + H⁺]

4-Amino-5-[3-amino-prop-1-ynyl]-7-[3-O-(2-cyanoethyl)-2-deoxy-β-D-*erythro*-pentofuranosyl)-7H-pyrrolo[2,3-d]pyrimidine

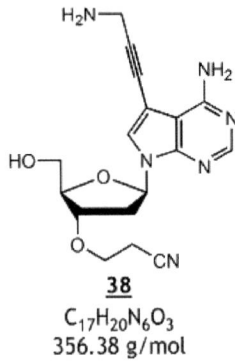

38
C₁₇H₂₀N₆O₃
356.38 g/mol

370 mg (0.82 mmol) of nucleoside **37** were dissolved in a mixture of 10 ml MeOH and 5 ml of 32 % aqueous ammonia. The pale yellow solution was stirred at room temperature overnight and concentrated after completion of the reaction. The brownish oily residue was put on a short silica gel column using CH₂Cl₂/MeOH (10 – 100 % MeOH) as eluent to give the product as glassy oil.

Yield: 220 mg (0.617 mmol, 75 %)
TLC: R_f = 0.06 (CH₂Cl₂/MeOH: 9/1)

¹H-NMR: (400 MHz, DMSO-d_6)
δ [ppm] = 8.1 (br s, 1H, H6), 7.69 (s, 1H, H2), 6.78 (br s, 2H, NH₂), 6.42 (dd, 1H, H1', J = 5.87 and J = 8.07), 5.21 (d, 1H, 5'-OH, J = 5.14), 4.25 (m, 1H, H3'), 3.97 (m, 1H, H4'), 3.67 (t, 2H, -O-CH₂, J = 6.26), 3.55 (m, 2H, H5'), 3.4 (br s, 2H, propargyl-CH₂), 2.81 (t, 2H, CH₂CN, J = 6.26), 2.64 – 2.34 (m, 2H, H2').

¹³C-NMR: (100 MHz, DMSO-d_6)
δ [ppm] = 157.4 (C4), 152.9 (C2), 149.3 (C7a), 125.9 (C6), 119.2 (CN), 102.1 (C4a), 95.5 (C5), 84.7 (C4'), 83.6 (propargyl-C-CH₂), 83.2 (C1'), 79.9 (C3'), 75.1

(propargyl-$\underline{C}$-C5), 63.4 (O-$\underline{C}$H$_2$), 61.9 (C5'), 36.5 (C2'), 30.7 (propargyl-$\underline{C}$H$_2$), 18.2 ($\underline{C}$H$_2$CN).

ESI(+)-MS: m/z 357.1 [M + H$^+$]

2-Amino-4-hydroxy-pyrrolo[2,3-d]pyrimidine

40
C$_6$H$_6$N$_4$O
150.14 g/mol

A mixture of 42.3 ml (0.26 mol) bromoacetaldehyde diethylacetal **20**, 6.8 ml 37 % hydrochloric acid and 137 ml water were heated up to 90 °C for 30 min with stirring. The clear pale yellow solution was allowed to cool down to room temperature and treated with 26.5 g (0.32 mol) sodium acetate. In another flask (500 ml), a suspension of 40.0 g (0.32 mol) 2,4-diamino-6-hydroxy-pyrimidine **39** and 13.3 g (0.16 mol) sodium acetate in 294 ml water was heated up to 70 – 85 °C. Having reached that temperature, the content from the first flask was added to the second one. With vigorous stirring, this mixture was heated up to 80 °C for 2 h within which the suspension turned into a solution first, followed by precipitation of purple crystalline product. The reaction mixture was cooled down to 0 °C for completion of precipitation. The crystals were filtered off, washed with each 100 ml cold water and acetone and dried under reduced pressure.

Yield: 26.8 g (178.5 mmol, 69 %)
TLC: R$_f$ = 0.55 (CH$_2$Cl$_2$/EtOH 2/1)

^{1}H-NMR: (400 MHz, DMSO-d_6)
     δ [ppm] = 10.96 (br s, 1H, H3), 10.26 (s, 1H, H7), 6.60 (dd, 1H, H6, J = 2.56 and J = 3.29), 6.18 (dd, 1H, H5, J = 2.56 and J = 3.29), 6.06 (br s, 2H, NH$_2$).

^{13}C-NMR: (100 MHz, DMSO-d_6)
     δ [ppm] = 159.0 (C4), 152.3 (C2), 151.2 (C7a), 116.7 (C6), 101.6 (C5), 99.9 (C4a).

ESI(+)-MS: m/z   150.7

2,2-Dimethyl-N-(4-hydroxy-pyrrolo[2,3-d]pyrimidin-2-yl)-propionamide

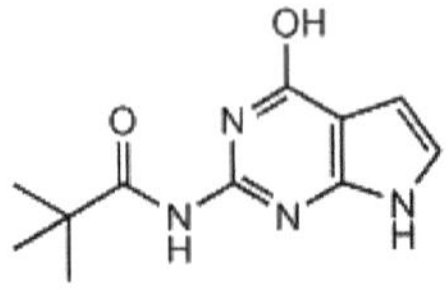

41
$C_{11}H_{14}N_4O_2$
234.25 g/mol

In a preheated flask, 4.54 g (30 mmol) of compound **40** and 13.1 ml (105 mmol) pivaloyl chloride were dissolved in 50 ml dry pyridine under an argon atmosphere. The violet suspension was heated up to 80 - 90 °C for about 30 min while it turned into a clear dark brown solution. The solvent was evaporated then, the residue dissolved in 30 ml MeOH and treated with 10 % aqueous ammonia until pH 4 causing the product precipitating as brown crystals. The crude product was filtered off, dried under reduced pressure and purified on column with CH_2Cl_2/MeOH (2 - 10 % MeOH).

Yield: 5.16 g (22 mmol, 73 %)
TLC: R_f = 0.3 (CH_2Cl_2/MeOH: 9/1)

^{1}H-NMR: (400 MHz, DMSO-d_6)
    δ [ppm] = 11.83 (s, 1H, piv-NH), 11.57 (s, 1H, H3), 10.79 (s, 1H, H7), 6.95 (m, 1H, H6), 6.40 (m, 1H, H5), 1.24 (s, 9H, CH_3).

^{13}C-NMR: (100 MHz, DMSO-d_6)
    δ [ppm] = 180.9 (piv-C=O), 157.0 (C4), 147.8 (C2), 146.5 (C7a), 119.7 (C6), 103.9 (C4a), 102.2 (C5), 40.3 (piv-$\underline{C}$-CH_3), 26.4 (piv-$\underline{C}H_3$).

ESI(+)-MS: m/z   234.8

2,2-Dimethyl-*N*-(4-hydroxy-5-iodo-pyrrolo[2,3-*d*]pyrimidin-2-yl)-propionamide

42

$C_{11}H_{13}IN_4O_2$
360.15 g/mol

1.0 g (4.3 mmol) of the heterocycle **41** was dissolved in 20 ml abs. DMF under an argon atmosphere. The solution was treated with 4 ml (15.4 mmol) of bis(trimethylsilyl)acetamide and stirred for 2 h at 40 °C. After cooling down to room temperature, 1.2 g (5.2 mmol) of *N*-iodosuccinimide were added in one portion. The solution was protected from light and stirred for another 2 h at room temperature. The dark brown solution was poured into 50 ml cold water with stirring. After 1-2 h the product precipitated as brown crystals which were filtered off, washed with 50 ml cold water and dried in vacuum.

Yield: 929 mg (2.58 mmol, 60 %)
TLC: R_f = 0.44 (CH_2Cl_2/MeOH: 9/1)

^{1}H-NMR: (400 MHz, DMSO-d_6)
δ [ppm] = 11.91 (br s, 1H, piv-NH), 11.81 (br s, 1H, H3), 10.83 (br s, 1H, H7), 7.15 (s, 1H, H6), 1.23 (s, 9H, CH_3).

^{13}C-NMR: (100 MHz, DMSO-d_6)
δ [ppm] = 180.9 (piv-C=O), 156.5 (C4), 147.9 (C2), 146.9 (C7a), 124.9 (C6), 103.8 (C4a), 54.2 (C5), 40.3 (piv-C̲-CH_3), 26.3 (piv-C̲H_3).

MALDI(+)-MS: *m/z* 361.81 [M + H$^+$]

2,2-Dimethyl-N-(4-chloro-pyrrolo[2,3-d]pyrimidin-2-yl)-propionamide

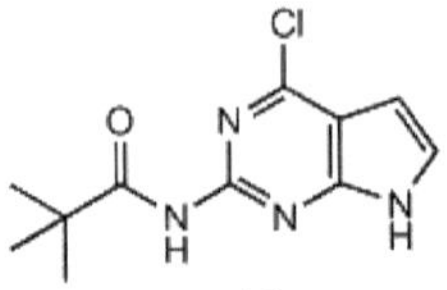

44
$C_{11}H_{13}ClN_4O$
252.70 g/mol

2.34 g (10 mmol) of the well-dried compound **41** were suspended under inert atmosphere in 50 ml $POCl_3$ and refluxed for about 2.5 h. Roughly 30 ml $POCl_3$ were then distilled off and the dark brown residue was poured slowly into 200 ml of ice water. The pH value of the reaction mixture was adjusted to 4 by dropwise addition of 10 % aqueous ammonia. The product precipitated as yellowish brown crystals which were collected by filtration, washed with cold water and dried in vacuum. The aqueous filtrate was extracted three times with 100 ml ethyl acetate each and the combined organic layers were washed once with brine, dried over $MgSO_4$ and concentrated to give a second fraction of the product.

Yield: 2.21 g (8.74 mmol, 87 %)
TLC: R_f = 0.65 (CH_2Cl_2/MeOH: 9/1)

^{1}H-NMR: (400 MHz, DMSO-d_6)
δ [ppm] = 12.34 (br s, 1H, piv-NH), 10.04 (br s, 1H, H7), 7.54 (d, 1H, H6), 6.53 (d, 1H, H5), 1.24 (s, 9H, CH_3).

^{13}C-NMR: (100 MHz, DMSO-d_6)
δ [ppm] = 176.0 (piv-C=O), 153.0 (C4), 151.6 (C2), 150.8 (C7a), 127.8 (C6), 113.3 (C4a), 99.1 (C5), 39.3 (piv-$\underline{C}$-CH_3), 27.2 (piv-$\underline{C}H_3$).

ESI(+)-MS: m/z 253.0

2,2-Dimethyl-N-(4-chloro-5-iodo-pyrrolo[2,3-d]pyrimidin-2-yl)-propionamide

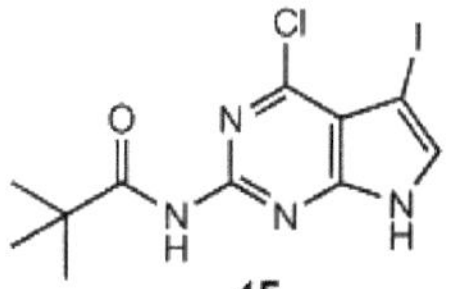

45

$C_{11}H_{12}ClIN_4O$
378.60 g/mol

Method A:

2.2 g (8.7 mmol) of compound **44** were dissolved in 40 ml abs. DMF and treated with 3.4 ml (13 mmol) of bis-(trimethylsilyl)acetamide. The mixture was stirred for 2 h at 40 °C under argon, then cooled down to room temperature and treated with 2.35 g (10.4 mmol) of N-iodosuccinimide. The orange solution was stirred for another 2 h in the dark at room temperature and poured into 100 ml cold water with stirring in order to quench the reaction. The aqueous layer was extracted three times with 150 ml CH_2Cl_2 each, the combined organic layers were washed once with 100 ml brine, dried over $MgSO_4$, filtered and concentrated. The crude residue was purified on a silica gel column with CH_2Cl_2/MeOH (2 - 10 % MeOH) as eluent to furnish the pure product as yellow crystals.

Yield: 2.50 g (6.6 mmol, 76 %)

Method B:

Under argon, 1.0 g (3.96 mmol) of the starting material **44** was dissolved in 12 ml dry DMF and treated with 1.12 g (4.95 mmol) of N-iodosuccinimide. The mixture was protected from light and stirred for 2 h at room temperature, diluted with 150 ml methylene chloride and washed with each 50 ml of brine and $Na_2S_2O_3$-solution. The oily residue obtained after solvent removal was purified on column chromatography (CH_2Cl_2/MeOH, 2 - 10 % MeOH) to furnish yellow crystalline product.

Yield: 1.39 g (3.67 mmol, 92 %)
TLC: R_f = 0.7 (CH_2Cl_2/MeOH: 9/1)

^{1}H-NMR: (400 MHz, DMSO-d_6)
δ [ppm] = 12.71 (br s, 1H, piv-NH), 10.12 (br s, 1H, H7), 7.77 (d, 1H, H6, J =

2.53), 1.23 (s, 9H, CH$_3$).

^{13}C-NMR: (100 MHz, DMSO-d_6)
δ [ppm] = 175.8 (pivaloyl-C=O), 152.5 (C4), 151.4 (C2), 150.5 (C7a), 132.7 (C6), 112.3 (C4a), 51.6 (C5), 39.7 (piv-C̲-CH$_3$), 26.8 (piv-C̲H$_3$).

ESI(+)-MS: m/z   378.8

4-Chloro-7-(2-deoxy-3,5-di-O-(4-toluoyl)-β-D-erythro-pentofuranosyl]-5-iodo-2-pivaloylamino-7H-pyrrolo[2,3-d]pyrimidine

46
C$_{32}$H$_{32}$ClIN$_4$O$_6$
730.98 g/mol

Under argon, 757 mg (2 mmol) of compound **45** were dissolved in 15 ml dry THF, treated with 90 mg (2.2 mmol) NaH (60 % in mineral oil) and the orange solution was stirred for 30 min at room temperature while it turned dark green. 933 mg (2.4 mmol) of compound **27** were added then and the cloudy mixture was stirred for another 1 h at room temperature within which it turned dark orange. The reaction was stopped by addition of Dowex (H$^+$-form), diluted in 100 ml CH$_2$Cl$_2$, filtered over Celite and concentrated under reduced pressure. The residual oil was purified on a silica gel column using CH$_2$Cl$_2$/MeOH (0 - 1 % MeOH) as eluent to give the product as yellowish crystals.

Yield: 628 mg (0.86 mmol, 43 %)
TLC: R$_f$ = 0.59 (CH$_2$Cl$_2$)

^{1}H-NMR: (400 MHz, CDCl$_3$)
δ [ppm] = 8.18 (br s, 1H, piv-NH), 8.02 - 7.28 (m, 8H, toluoyl-H), 7.45 (s, 1H, H6), 6.76 (m, 1H, H1'), 5.82 (m, 1H, H3'), 4.74 (m, 2H, H5'), 4.61 (m, 1H, H4'), 2.98 - 2.86 (m, 2H, H2'), 2.47 (s, 3H, toluoyl-CH$_3$), 2.46 (s, 3H, toluoyl-CH$_3$),

1.38 (s, 9H, piv-CH₃).

¹³C-NMR: (100 MHz, CDCl₃)
δ [ppm] = 175.7 (pivaloyl-C=O), 166.3 (toluoyl-C=O), 166.2 (toluoyl-C=O), 153 (C2), 151.5 (C4), 144.5 (toluoyl-C), 144.3 (toluoyl-C), 131.1 (C6), 130 (toluoyl-C), 129.7 (toluoyl-C), 129.5 (toluoyl-C), 129.4 (toluoyl-C), 126.9 (toluoyl-C), 126.6 (C7a), 113.7 (C4a), 84.7 (C1'), 82.8 (C4'), 75.1 (C3'), 64.1 (C5'), 53.0 (C5), 40.4 (piv-C-CH₃), 38.2 (C2'), 27.5 (piv-CH₃), 21.8 (toluoyl-CH₃).

ESI(+)-MS: m/z   731.4 [M + H⁺]

4-Chloro-2-methylthio-pyrrolo[2,3-d]pyrimidine

47
$C_{11}H_{12}ClN_4O$
199.66 g/mol

1.0 g (5.5 mmol) of heterocycle **16** were suspended in 30 ml of phosphorus oxychloride and heated up to reflux for 4 h while the starting material completely dissolved. Afterwards the mixture was allowed to cool down to room temperature and quenched on 200 ml ice water with stirring. By addition of aqueous ammonia the pH value was adjusted to 4 causing product precipitation. The crystalline product was filtered off then and the aqueous layer was extracted three times with 200 ml ethyl acetate each. The combined organic layers were dried over MgSO₄ and evaporated to dryness to give a second fraction of yellow crystalline product. Both crystalline fractions were re-crystallized in ethyl acetate, filtered off and dried in vacuum.

Yield: 400 mg (2 mmol, 36 %)
TLC: R_f = 0.23 (n-hexane/ethyl acetate: 4/1)

¹H-NMR: (400 MHz, DMSO-d₆)
δ [ppm] = 12.39 (br s, 1H, H7), 7.51 (dd, 1H, H6, J = 3.54 and J = 2.27), 6.51 (dd, 1H, H5, J = 3.54 and J = 1.77), 2.55 (s, 9H, CH₃).

¹³C-NMR: (100 MHz, DMSO-d₆)

δ [ppm] = 162.7 (C2), 152.7 (C4), 150.4 (C7a), 126.8 (C6), 113.2 (C4a), 99.0 (C5), 13.8 (SMe).

ESI(+)-MS: m/z   199.6

4-Chloro-5-iodo-2-methylthio-pyrrolo[2,3-d]pyrimidine

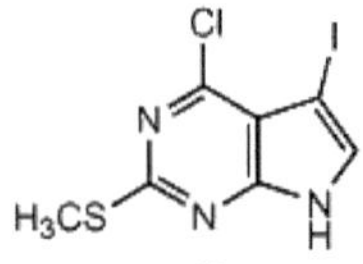

48

$C_7H_5ClIN_3S$
325.56 g/mol

Under argon, 242 mg (1.2 mmol) of compound **47** were dissolved in 50 ml dry methylene chloride and treated with 300 mg (1.32 mmol) N-iodosuccinimide. The reaction mixture was stirred for 3 h until completion of the reaction at room temperature. The solvent was removed under reduced pressure and the yellowish-crystalline product obtained after purification on column chromato-graphy with CH_2Cl_2/MeOH (0 - 2 % MeOH) as eluent.

Yield: 260 mg (0.8 mmol, 67 %)
TLC: R_f = 0.57 (CH_2Cl_2/MeOH: 9/1)

^{1}H-NMR (250 MHz, DMSO-d_6)
δ [ppm] = 12.73 (br s, 1H, H7), 7.75 (d, 1H, H6), 2.54 (s, 3H, SCH$_3$).

^{13}C-NMR: (63 MHz, DMSO-d_6)
δ [ppm] = 163.2 (C2), 152.4 (C4), 150.7 (C7a), 132.3 (C6), 112.5 (C4a), 52.1 (C5), 13.7 (SMe).

ESI(+)-MS: m/z   325.8

4-Chloro-7-[2-deoxy-3,5-di-O-(4-toluoyl)-β-D-*erythro*-pentofuranosyl]-2-methylthio-7*H*-pyrrolo[2,3-*d*]pyrimidine

49
$C_{28}H_{26}ClN_3O_5S$
552.04 g/mol

210 mg (1.05 mmol) of the well-dried heterocycle **47** were dissolved in 30 ml dry acetonitrile and treated with 50 mg (1.16 mmol) of 60 % sodium hydride in mineral oil. The yellow solution was stirred under argon for 30 min at room temperature, then 408 mg (1.05 mmol) of compound **27** were added in one lot. The cloudy suspension was stirred for 3 h at room temperature for completion of the reaction and quenched by the addition of Dowex (H$^+$-form) then. After filtration through Celite, the filtrate was concentrated and purified on column chromatography with *n*-hexane/ethyl acetate 4/1 to give the product as colorless foam.

Yield: 419 mg (0.76 mmol, 72 %)
TLC: R$_f$ = 0.36 (*n*-hexane/ethyl acetate: 4/1)

^{1}H-NMR: (250 MHz, CDCl$_3$)
δ [ppm] = 7.94 (m, 4H, toluoyl-H$_A$, *J* = 8.23), 7.26 (m, 4H, toluoyl-H$_B$, *J* = 6.86), 7.24 (d, 1H, H6), 6.74 (dd, 1H, H1′, *J* = 6.04 and *J* = 7.96), 6.49 (d, 1H, H5, *J* = 3.84), 5.75 (m, 1H, H3′), 4.75 – 4.53 (m, 3H, H4′, H5′), 2.89 (m, 1H, H2′), 2.75 (m, 1H, H2′), 2.64 (s, 3H, SCH$_3$), 2.44 (s, 3H, toluoyl-CH$_3$), 2.42 (s, 3H, toluoyl-CH$_3$).

^{13}C-NMR: (63 MHz, CDCl$_3$)
δ [ppm] = 166.3 (C=O), 166.1 (C=O), 165 (C2), 152.2 (C4), 152.1 (C7a), 144.6 (toluoyl-C$_{quart}$), 144.3 (toluoyl-C$_{quart}$), 129.9 (toluoyl-C$_A$), 129.7 (toluoyl-C$_A$),

129.4 (toluoyl-C$_B$), 129.2 (toluoyl-C$_B$), 126.8 (toluoyl-C$_C$), 126.6 (toluoyl-C$_C$), 124.4 (C6), 114.8 (C4a), 101.4 (C5), 84.3 (C1'), 82.4 (C4'), 75.1 (C3'), 64.3 (C5'), 38.2 (C2'), 21.8 (toluoyl-CH$_3$), 21.8 (toluoyl-CH$_3$).

ESI(+)-MS: m/z   552.2

4-Chloro-7-[2-deoxy-3,5-di-O-(4-toluoyl)-β-D-*erythro*-pentofuranosyl]-5-iodo-2-methylthio-*7H*-pyrrolo[2,3-*d*]pyrimidine

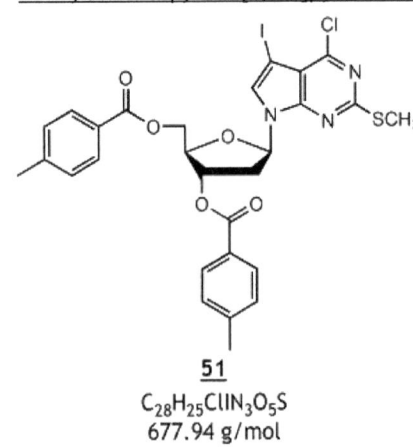

51
C$_{28}$H$_{25}$ClIN$_3$O$_5$S
677.94 g/mol

120 mg (0.37 mmol) of the heterocycle **48** were dissolved with heating at 60 °C in 20 ml dry acetonitrile within 15 min. The solution was cooled down to approximately 30 °C and 17 mg (0.4 mmol) of sodium hydride (60 % in mineral oil) were added in one lot. The mixture was stirred for 15 min at room temperature, followed by subsequent addition of 143 mg (0.37 mmol) of compound **27**. The yellow solution was stirred for 1 h at room temperature and the reaction stopped by addition of Dowex (H$^+$-form). The mixture was diluted in 50 ml methylene chloride and filtered through Celite. The residue obtained after concentration of the orange clear filtrate was placed atop of a silica gel column for purification using *n*-hexane/ethyl acetate 4/1 to give the product as colorless crystals.

Yield: 137 mg (0.2 mmol, 55 %)
TLC: R$_f$ = 0.34 (*n*-hexane/ethyl acetate: 4/1)

^{1}H-NMR: (400 MHz, CDCl$_3$)
δ [ppm] = 7.97 (d, 2H, toluoyl-H$_A$, J = 8.08), 7.91 (d, 2H, toluoyl-H$_A$, J = 8.08), 7.38 (s, 1H, H6), 7.28 (t, 4H, toluoyl-H$_B$, J = 8.34), 6.71 (t, 1H, H1', J = 6.82),

5.72 (m, 1H, H3'), 4.69 (qd, 2H, H5', J = 3.54, J = 12.13 and J = 33.09), 4.59 (m, 1H, H4'), 2.76 (m, 2H, H2'), 2.62 (s, 3H, SCH$_3$) 2.44 (s, 3H, toluoyl-CH$_3$), 2.43 (s, 3H, toluoyl-CH$_3$).

^{13}C-NMR: (100 MHz, CDCl$_3$)
δ [ppm] = 166.3 (toluoyl-C=O), 166.1 (toluoyl-C=O), 165.2 (C2), 153.8 (C4), 152.7 (C7a), 144.4 (toluoyl-C$_{quart}$), 129.9 (C6), 129.8 (toluoyl-C$_A$), 129.5 (toluoyl-C$_A$), 129.3 (toluoyl-C$_B$), 126.7 (toluoyl-C$_C$), 126.6 (toluoyl-C$_C$), 114.8 (C4a), 84.3 (C1'), 82.8 (C4'), 75.1 (C3'), 64.1 (C5'), 53.4 (C5), 38.7 (C2'), 21.9 (toluoyl-CH$_3$), 21.8 (toluoyl-CH$_3$), 14.7 (SCH$_3$).

ESI(+)-MS: m/z   678.0

7-[2-Deoxy-3,5-di-O-(4-toluoyl)-β-D-*erythro*-pentofuranosyl]-5-iodo-2-methylthio-*7H*-pyrrolo[2,3-*d*]pyrimidin-4-one

52
C$_{28}$H$_{26}$IN$_3$O$_6$S
659.49 g/mol

Under argon, 103 mg (0.15 mmol) of nucleoside **51** were dissolved in 4 ml dry DMF and 2.7 ml 1,4-dioxane. To this solution, 92 mg (0.75 mmol) of *syn*-pyridinealdoxime and 105 µl (0.83 mmol) 1,1,3,3-tetramethylguanidine were added and the mixture was stirred for 24 h at room temperature. After removal of the solvent, the residue was taken up in 100 ml methylene chloride, washed with 0.1 M aqueous citric acid (2 x 50 ml), water (100 ml) and saturated aqueous NaHCO$_3$ (100 ml), then dried over MgSO$_4$ and concentrated after filtration. The yellow crystalline product was purified on a silica gel column using a gradient starting with CH$_2$Cl$_2$/ethyl acetate 95/5 to CH$_2$Cl$_2$/MeOH 95/5.

Yield: 88 mg (0.13 mmol, 89 %)
TLC: R_f = 0.39 (CH$_2$Cl$_2$/MeOH: 96/4)

^{1}H-NMR: (400 MHz, CDCl$_3$)
δ [ppm] = 10.76 (br s, 1H, NH), 7.94 (t, 4H, toluoyl-H$_A$, J = 8.34), 7.27 (d, 4H, toluoyl-H$_B$, J = 8.08), 7.05 (s, 1H, H6), 6.62 (dd, 1H, H1', J = 6.32 and J = 7.83), 5.69 (m, 1H, H3'), 4.67 (m, 2H, H5'), 4.56 (q, 1H, H4', J = 3.03), 2.72 (m, 2H, H2'), 2.64 (s, 3H, SCH$_3$), 2.44 (s, 3H, CH$_3$), 2.42 (s, 3H, CH$_3$).

^{13}C-NMR: (100 MHz, CDCl$_3$)
δ [ppm] = 166.5 (toluoyl-C=O), 165.9 (toluoyl-C=O), 159 (C4), 156.5 (C2), 147.2 (C7a), 144.5 (toluoyl-C$_{quart}$), 129.9 (toluoyl-C$_A$), 129.8 (toluoyl-C$_A$), 129.6 (toluoyl-C$_B$), 129.4 (toluoyl-C$_B$), 126.7 (toluoyl-C$_C$), 123.8 (C6), 105.8 (C4a), 84.0 (C1'), 82.6 (C4'), 75 (C3'), 64.1 (C5'), 55.8 (C5), 38.7 (C2'), 21.9 (toluoyl-CH$_3$), 21.8 (toluoyl-CH$_3$), 13.9 (SCH$_3$).

ESI(+)-MS: m/z 660.1 [M + H$^+$]

2-Amino-7-[2-deoxy-β-D-*erythro*-pentofuranosyl]-5-iodo-7*H*-pyrrolo[2,3-*d*]pyrimidin-4-one

54
C$_{11}$H$_{13}$IN$_4$O$_4$
392.15 g/mol

220 mg (0.33 mmol) of compound **52** were dissolved in 20 ml methylene chloride and cooled down to 0 °C. 136 mg (0.43 mmol) of 3-chloroperoxy-benzoic acid (50 – 60 % MCPBA) were added in one lot and the mixture was stirred for 15 min at 0 °C, then for 105 min at room temperature. The solvent was evaporated and the residue quickly purified on a short silica gel column using CH$_2$Cl$_2$/MeOH 96/4 as eluent. The oxidized nucleoside was suspended in 100 ml of a sat. NH$_4$OH/dioxane solution and filled into a sealable Parr bomb. The mixture was heated for 15 h at 110 °C, cooled and concentrated. The brownish oily residue was taken up in 100 ml CH$_2$Cl$_2$, washed with 100

ml sat. aq. NaHCO₃, dried over MgSO₄ and concentrated. The resulting yellow-brownish oil was suspended in 100 ml of a sat. NH₄OH/MeOH solution and filled in the Parr autoclave again. The mixture was heated up to 140 °C for 2 h, cooled down and concentrated. The oily residue was purified on a silica gel column using CH₂Cl₂/MeOH 9/1 yielding the product as yellow crystals.

Yield: 25 mg (0.06 mmol, 19 %)
TLC: R_f = 0.13 (CH₂Cl₂/MeOH: 9/1)

¹H-NMR: (400 MHz, DMSO-d_6)
δ [ppm] = 10.4 (br s, 1H, NH3), 7.11 (s, 1H, H6), 6.32 (br s, 2H, NH₂), 6.26 (dd, 1H, H1', J = 5.81 and J = 8.59), 5.19 (d, 1H, 3'-OH, J = 3.54), 4.89 (t, 1H, 5'-OH, J = 5.31), 4.26 (m, 1H, H3'), 3.74 (m, 1H, H4'), 3.48 (m, 2H, H5'), 2.31 (m, 1H, H2'), 2.04 (m, 1H, H2').

¹³C-NMR: (100 MHz, DMSO-d_6)
δ [ppm] = 158.1 (C4), 152.7 (C2), 150.6 (C7a), 121.7 (C6), 99.8 (C4a), 87.1 (C4'), 82.2 (C1'), 70.9 (C3'), 61.9 (C5'), 55 (C5), 39.8 (C2').

ESI(-)-MS: m/z 390.9 [M - H⁺]

7-[2-Deoxy-3,5-O-(tetraisopropyldisiloxane-1,3-diyl)-β-D-*erythro*-pentofuranosyl]-2N-(*N,N*-dimethylaminomethylidenyl)-5-iodo-7*H*-pyrrolo[2,3-*d*]pyrimidin-4-one

56
C₂₆H₄₄IN₅O₅Si₂
689.73 g/mol

1.0 g (2.55 mmol) of nucleoside **54** was coevaporated three times with pyridine, then dissolved in 10 ml dry pyridine and cooled down to 0 °C. 0.88 ml (2.81 mmol) of 1,1,3,3-tetraisopropyldichlordisiloxane were added and the mixture was stirred under argon 30 min at 0 °C first, then for 20 h at room temperature. The reaction was stopped by addition of 10 ml MeOH and stirring for 15 min. The solvents were removed, the oily residue was dissolved in 15 ml dry *N,N*-dimethylformamide again and treated with 1.7 ml (12.8 mmol) *N,N*-dimethylformamide dimethylacetal. The yellow solution was stirred for 24 h at room temperature under argon while the reaction was monitored by TLC. The mixture was concentrated and the crude product purified on a silica gel column using CH_2Cl_2/MeOH (5 - 10 % MeOH) as eluent.

Yield: 1.73 g (2.51 mmol, 98 %)
TLC: R_f = 0.2 (CH_2Cl_2/MeOH: 95/5)

^{1}H-NMR: (250 MHz, DMSO-d_6)
 δ [ppm] = 11.11 (br s, 1H, NH3), 8.57 (s, 1H, formamidino-CH), 7.12 (s, 1H, H6), 6.36 (dd, 1H, H1', *J* = 3.91 and *J* = 7.33), 4.67 (m, 1H, H3'), 3.9 (m, 2H, H5'), 3.71 (m, 1H, H4'), 3.14 (s, 3H, formamidino-CH$_3$), 3.02 (s, 3H, formamidino-CH$_3$), 2.59 - 2.36 (m, 2H, H2'), 1.02 (m, 28H, *iso*-propyl-CH$_3$).

^{13}C-NMR: (63 MHz, DMSO-d_6)
 δ [ppm] = 158.9 (C4), 157.7 (formamidino-C=N), 156.3 (C2), 149.2 (C7a), 122.6 (C6), 102.7 (C4a), 84 (C4'), 80.4 (C1'), 70.6 (C3'), 61.9 (C5'), 55.6 (C5), 40.6 (formamidino-CH$_3$), 40.2 (C2'), 34.5 (formamidino-CH$_3$), 17.3 - 16.7 (*iso*-propyl-CH$_3$), 12.7 - 12 (*iso*-propyl-$\underline{C}$H-CH$_3$).

ESI(+)-MS: *m/z* 690.3 [M + H$^+$]

<u>3N-Benzoyl-7-[2-deoxy-3,5-*O*-(tetraisopropyldisiloxane-1,3-diyl)-β-D-*erythro*-pentofuranosyl]-2N-(*N,N*-dimethylaminomethylidenyl)-5-iodo-7*H*-pyrrolo[2,3-*d*]pyrimidin-4-one</u>

57
$C_{33}H_{48}IN_5O_6Si_2$
793.84 g/mol

1.59 g (2.3 mmol) of nucleoside **56** were coevaporated three times with dry pyridine, then dissolved in 15 ml abs. pyridine and cooled down to 0 °C. 402 μl (3.45 mmol) of benzoyl chloride, diluted in 0.5 ml dry CH_2Cl_2, were added and the mixture was stirred for 30 min at 0 °C, then for 5 h at room temperature. After treatment with a second portion of benzoyl chloride (134 µl, 1.15 mmol) the reaction was allowed to proceed overnight. The mixture was concentrated, the residue taken up in 200 ml CH_2Cl_2 and washed with each 50 ml of water, sat. aqueous $NaHCO_3$ and brine. The organic layer was dried over $MgSO_4$, filtered and concentrated. The residue was purified on silica gel using a gradient starting with n-hexane/ethyl acetate = 5/1 to CH_2Cl_2/MeOH 95/5.

Yield: 1.4 g (1.76 mmol, 77 %)
TLC: R_f = 0.61 (CH_2Cl_2/MeOH: 95/5)

^{1}H-NMR: (250 MHz, $CDCl_3$)
δ [ppm] = 8.49 (s, 1H, formamidino-H), 7.85 (ψd, 2H, benzoyl-H_A, J = 7.34), 7.55 (ψt, 1H, benzoyl-H_C, J = 7.34), 7.41 (ψt, 2H, benzoyl-H_B, J = 7.83), 7.0 (s, 1H, H6), 6.47 (dd, 1H, H1', J = 3.42 and J = 7.34), 4.71 (q, 1H, H3', J = 7.58), 4.02 (m, 2H, H5'), 3.82 (m, 1H, H4'), 3.06 (s, 3H, formamidino-CH_3), 2.69 (s, 3H, formamidino-CH_3), 2.53 (m, 2H, H2'), 1.16 - 1.04 (m, 28H, iso-propyl-CH).

^{13}C-NMR: (63 MHz, $CDCl_3$)
δ [ppm] = 171.4 (benzoyl-C=O), 158.8 (C4), 156.3 (formamidino-C=N), 154.3 (C2), 148.3 (C7a), 133.9 (benzoyl-C_C), 132.8 (benzoyl-C_{quart}), 130.3 (benzoyl-C_A), 128.9 (benzoyl-C_B), 123.1 (C6), 103.9 (C4a), 84.9 (C4'), 81.2 (C1'), 70.2 (C3'), 61.9 (C5'), 55.8 (C5), 41.2 (formamidino-CH_3), 38.4 (C2'), 35.1 (formamidino-CH_3), 17.7 - 17.1 (iso-propyl-CH_3), 13.6 – 12.7 (iso-propyl-$\underline{C}H$-CH_3).

ESI(+)-MS: m/z 794.5 [M + H$^+$]

7-[2-Deoxy-β-D-erythro-pentofuranosyl]-2N-(N,N-dimethylaminomethylidenyl)-5-iodo 7H-pyrrolo[2,3-d]pyrimidin-4-one

58
$C_{21}H_{22}IN_5O_5$
551.33 g/mol

397 mg (0.5 mmol) **57** were dissolved in 10 ml dry THF and treated with 285 µl (1.75 mmol) of a triethylamine trihydrofluoride solution (approx. 37 % HF). The mixture was stirred for 4 h at room temperature, then concentrated and the residue purified on a short silica gel column with CH_2Cl_2/MeOH 9/1 to give the product as yellowish glassy crystals.

Yield: 276 mg (0.5 mmol, 100 %)
TLC: R_f = 0.27 (CH_2Cl_2/MeOH: 9/1)

^{1}H-NMR: (250 MHz, DMSO-d_6)
δ [ppm] = 8.58 (s, 1H, formamidino-CH), 7.80 (ψd, 2H, benzoyl-H_A, J = 7.09 and J = 8.56), 7.7 (ψt, 1H, benzoyl-H_C, J = 7.34), 7.55 (ψt, 2H, benzoyl-H_B, J = 7.34), 7.4 (s, 1H, H6), 6.47 (dd, 1H, H1′, J = 5.87 and J = 7.58), 5.29 (d, 1H, 3′-OH, J = 3.91), 4.93 (t, 1H, 5′-OH, J = 5.14), 4.34 (m, 1H, H3′), 3.81 (m, 1H, H4′), 3.54 (m, 2H, H5′), 3.09 (s, 3H, formamidino-CH_3), 2.61 (s, 3H, formamidino-CH_3), 2.44 – 2.13 (m, 2H, H2′).

^{13}C-NMR: (63 MHz, DMSO-d_6)
δ [ppm] = 171.7 (benzoyl-C=O), 157.8 (C4), 156.5 (formamidino-C=N), 153.8 (C2), 148.7 (C7a), 134.3 (benzoyl-C_C), 133.1 (benzoyl-C_{quart}), 129.6 (benzoyl-C_A), 129.2 (benzoyl-C_B), 124.2 (C6), 102.1 (C4a), 87.4 (C4′), 82.4 (C1′), 70.9 (C3′), 61.8 (C5′), 55.7 (C5), 40.6 (formamidino-CH_3), 39.9 (C2′), 34.4 (formamidino-CH_3).

ESI(+)-MS: m/z 552.0 [M + H$^+$]

3N-Benzoyl-7-[2-deoxy-5-O-(4-monomethoxytrityl)-β-D-*erythro*-pentofuranosyl]-2N-(*N,N*-dimethylaminomethylidenyl)-5-iodo-7*H*-pyrrolo[2,3-*d*]pyrimidin-4-one

59
$C_{41}H_{38}IN_5O_6$
823.67 g/mol

845 mg (1.53 mmol) of the starting material **58** were coevaporated three times with

pyridine and dissolved in 10 ml abs. pyridine. 19 mg (0.15 mmol) of 4-dimethylaminopyridine and 617 mg (2 mmol) p-monomethoxytrityl chloride were added and the mixture was stirred under argon overnight at room temperature. The reaction was stopped by addition of 2 ml MeOH, then concentrated and the residue taken up in 100 ml CH_2Cl_2. The organic layer was washed with 50 ml water, sat. aq. $NaHCO_3$ and brine each, dried over $MgSO_4$, filtered and concentrated. The purification of the crude material on silica gel column with CH_2Cl_2/MeOH (0 - 5 % MeOH) gave the product as yellow foam.

Yield: 1.04 g (1.26 mmol, 82 %)
TLC: R_f = 0.42 (CH_2Cl_2/MeOH: 95/5)

^{1}H-NMR: (400 MHz, $CDCl_3$)
δ [ppm] = 8.51 (s, 1H, formamidino-CH), 7.87 (ψd, 2H, benzoyl-H_A, J = 7.68), 7.55 (ψt, 1H, benzoyl-H_C, J = 7.32), 7.46 - 7.23 (m, 14H, benzoyl-H_B, J = 7.68, MMTr-H), 6.93 (s, 1H, H6), 6.86 (ψd, 2H, MMTr-H, J = 8.78), 6.60 (ψt, 1H, H1', J = 6.22), 4.56 (m, 1H, H3'), 4.07 (m, 1H, H4'), 3.81 (s, 3H, OCH_3), 3.47 - 3.27 (m, 2H, H5'), 3.06 (s, 3H, formami-dino-CH_3), 2.69 (s, 3H, formamidino-CH_3), 2.50 - 2.36 (m, 2H, H2').

^{13}C-NMR: (100 MHz, $CDCl_3$)
δ [ppm] = 171.4 (benzoyl-C=O), 158.6 (MMT-$\underline{C}$-OMe), 158 (C4), 157.7 (MMT-C), 156.4 (formamidino-C=N), 153.7 (C2), 149.4 (MMT-C), 148.5 (C7a), 144.1 (MMT-C), 144 (MMT-C), 134.8 (MMT-C), 134.1 (benzoyl-C_C), 132.9 (benzoyl-C_{quart}), 129.8 (benzoyl-C_A), 129.4 (benzoyl-C_B), 129 (MMT-C), 127.9 (MMT-C), 127.8 (MMT-C), 127.7 (MMT-C), 126.7 (MMT-C), 123 (C6), 113.5 (MMT-C), 102.1 (C4a), 85.8 (C4'), 82.3 (C1'), 72.9 (C3'), 64.3 (C5'), 55.7 (C5), 55 (MMT-O$\underline{C}H_3$), 40.5 (formamidino-CH_3), 39.9 (C2'), 34.3 (formamidino-CH_3).

ESI(+)-MS: m/z   824.7 [M + H$^+$]

3N-Benzoyl-7-[3-O-(2-cyanoethyl)-2-deoxy-5-O-(4-monomethoxytrityl)-β-D-*erythro*-pentofuranosyl]-2N-(*N,N*-dimethylaminomethylidenyl)-5-iodo-7H-pyrrolo[2,3-*d*]pyrimidin-4-one

60
$C_{44}H_{41}IN_6O_6$
876.74 g/mol

In an Erlenmeyer flask with magnetic stirrer and triangle stirring bar, 1.02 g (1.25 mmol) of compound **58** were dissolved in 12 ml *tert*-butanol and 3.3 ml (50 mmol) freshly distilled acrylonitrile under argon. The mixture was agitated vigorously for a few minutes at room temperature, then 407 mg (1.25 mmol) of cesium carbonate were added and vigorous stirring continued for 4 h. The suspension was taken up in 200 ml methylene chloride then, filtered through Celite and concentrated. The residue was purified on a silica gel column with CH_2Cl_2 as eluent. The product was obtained as yellow foam.

Yield: 759 mg (0.87 mmol, 69 %)
TLC: R_f = 0.69 (CH_2Cl_2/MeOH: 95/5)

^{1}H-NMR: (400 MHz, $CDCl_3$)
δ [ppm] = 8.54 (s, 1H, formamidino-CH), 7.87 (ψd, 2H, benzoyl-H_A, *J* = 7.32), 7.55 (t, 1H, benzoyl-H_C, *J* = 7.32), 7.44 (t, 2H, benzoyl-H_B, *J* = 7.68), 7.39 - 7.23 (m, 12H, MMT), 6.95 (s, 1H, H6), 6.87 (ψd, 2H, MMT, *J* = 8.78), 6.56 (dd, 1H, H1', *J* = 5.86 and *J* = 8.78), 4.18 (m, 2H, H3', H4'), 3.81 (s, 3H, OCH_3), 3.68 (t, 2H, -O-C$\underline{H}_2$, *J* = 6.22), 3.41 - 3.29 (m, 2H, H5'), 3.06 (s, 3H, formamidino-CH_3), 2.69 (s, 3H, formami-dino-CH_3), 2.62 (t, 2H, C$\underline{H}_2$CN, *J* = 6.22), 2.48 - 2.31 (m, 2H, H2').

^{13}C-NMR: (100 MHz, $CDCl_3$)
δ [ppm] = 171.4 (benzoyl-C=O), 158.1 (C4), 157.7 (MMT-C), 156.5 (formami-dino-C=N), 153.8 (C2), 148.7 (C7a), 144.1 (MMT-C), 144 (MMT-C), 134.8 (MMT-C), 134.1 (benzoyl-C_C), 132.9 (benzoyl-C_{quart}), 130.2 (MMT-C), 129.9 (benzoyl-

C_A), 129.4 (benzoyl-C_B), 129 (MMT-C), 127.9 (MMT-C), 127.8 (MMT-C), 126.8 (MMT-C), 123.8 (C6), 119 (CN), 113.5 (MMT-C), 102.2 (C4a), 82.8 (C4'), 82.2 (C1'), 80.7 (C3'), 63.8 (C5'), 63.6 (-O-$\underline{C}$H$_2$), 55.9 (C5), 55.1 (MMT-O$\underline{C}$H$_3$), 40.8 (formamidino-CH$_3$), 36.7 (C2'), 34.7 (formamidino-CH$_3$), 18.9 ($\underline{C}$H$_2$CN).

ESI(+)-MS: m/z 877.5 [M + H$^+$]

3N-Benzoyl-7-[3-O-(2-cyanoethyl)-2-deoxy-β-D-*erythro*-pentofuranosyl]-2N-(N,N-dimethylaminomethylidenyl)-5-iodo-7H-pyrrolo[2,3-*d*]pyrimidin-4-one

61
C$_{24}$H$_{25}$IN$_6$O$_5$
604.40 g/mol

740 mg (0.84 mmol) of compound **60** were dissolved in 10 ml CH$_2$Cl$_2$/EtOH (1/1) and treated with 321 mg (1.69 mmol) *p*-toluenesulfonic acid monohydrate. The mixture was stirred for 1.5 h, then concentrated and the residue taken up in 100 ml CH$_2$Cl$_2$. The organic layer was washed with each 50 ml of aq. sat. NaHCO$_3$, brine and water, dried over MgSO$_4$ and filtered. After solvent removal the crude product was purified on column chromatography using CH$_2$Cl$_2$/MeOH 98/2 as eluent to give the pure compound as yellow foam.

Yield: 386 mg (0.64 mmol, 76 %)
TLC: R$_f$ = 0.22 (CH$_2$Cl$_2$/MeOH: 95/5)

^{1}H-NMR: (250 MHz, DMSO-d_6)
δ [ppm] = 8.57 (s, 1H, formamidino-CH), 7.78 (ψd, 2H, benzoyl-H$_A$, J = 7.09 and J = 8.31), 7.70 (ψt, 1H, benzoyl-H$_C$, J = 7.58), 7.55 (ψt, 2H, benzoyl-H$_B$, J = 7.58), 7.42 (s, 1H, H6), 6.43 (dd, 1H, H1', J = 6.11 and J = 8.07), 5.1 (t, 1H, 5'-OH, J = 5.14), 4.24 (m, 1H, H3'), 3.95 (m, 1H, H4'), 3.69 (t, 2H, -O-C$\underline{H}_2$, J = 6.11), 3.56 (m, 2H, H5'), 3.07 (s, 3H, formamidino-CH$_3$), 2.8 (t, 2H, C$\underline{H}_2$CN, J = 6.11), 2.61 (s, 3H, formamidino-CH$_3$), 2.42 (m, 2H, H2').

¹³C-NMR: (63 MHz, DMSO-d_6)
δ [ppm] = 171.7 (benzoyl-C=O), 158 (C4), 156.7 (formamidino-C=N), 154 (C2), 148.9 (C7a), 134.5 (benzoyl-C_C), 133.1 (benzoyl-C_{quart}), 129.7 (benzoyl-C_A), 129.3 (benzoyl-C_B), 124.2 (C6), 119.5 (CN), 102.2 (C4a), 84.8 (C4'), 82.5 (C1'), 79.9 (C3'), 63.5 (-O-$\underline{C}H_2$), 61.9 (C5'), 56 (C5), 40.7 (formamidino-CH₃), 37.1 (C2'), 34.5 (formamidino-CH₃), 18.4 ($\underline{C}H_2CN$).

ESI(+)-MS: m/z   605.2 [M + H⁺]

7-[3-O-(2-Cyanoethyl)-2-deoxy-5-O-β-D-*erythro*-pentofuranosyl]-5-iodo-7H-pyrrolo[2,3-d]pyrimidin-4-one

62
$C_{14}H_{16}IN_5O_4$
445.21 g/mol

349 mg (0.58 mmol) of nucleoside **61** were dissolved in 10 ml MeOH and 10 ml of 32 % aq. ammonia and stirred at room temperature for 24 h. After completion of the reaction, the solvents were evaporated and the residue was purified on column chromatography using CH₂Cl₂/MeOH 9/1 to give the product as yellowish crystals.

Yield: 187 mg (0.42 mmol, 72 %)
TLC: R_f = 0.25 (CH₂Cl₂/MeOH: 9/1)

¹H-NMR: (400 MHz, DMSO-d_6)
δ [ppm] = 10.46 (br s, 1H, NH3), 7.14 (s, 1H, H6), 6.36 (br s, 2H, NH₂), 6.22 (dd, 1H, H1', J = 5.87 and J = 8.80), 5.02 (t, 1H, 5'-OH, J = 5.38), 4.16 (m, 1H, H3'), 3.9 (m, 1H, H4'), 3.65 (t, 2H, -O-$\underline{C}H_2$, J = 6.11), 3.51 (m, 2H, H5'), 2.78 (t, 2H, $\underline{C}H_2CN$, J = 6.11), 2.45 - 2.22 (m, 2H, H2').

¹³C-NMR: (100 MHz, DMSO-d_6)
δ [ppm] = 158.1 (C4), 152.8 (C2), 150.8 (C7a), 121.7 (C6), 119.4 ($\underline{C}N$), 99.9 (C4a), 87 (C4'), 82.2 (C1'), 80.1 (C3'), 63.5 (-O-$\underline{C}H_2$), 61.9 (C5'), 55.3 (C5), 39.9 (C2'), 18.3 ($\underline{C}H_2CN$).

ESI(+)-MS: m/z   446.0 [M + H⁺]

<u>2'-Deoxy-(3',5'-O-diacetyl)cytidine</u>

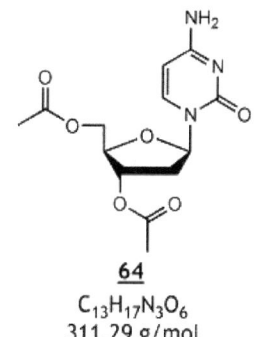

64
$C_{13}H_{17}N_3O_6$
311.29 g/mol

10 g (37.9 mmol) of 2'-deoxycytidine hydrochloride **63** were suspended in 130 ml glacial acetic acid and 10.8 ml (152 mmol) of acetyl chloride were added dropwise. The suspension was agitated overnight at room temperature, the solvent was distilled off and the solid residue recrystallized in EtOH to give the product as colorless crystals.

Yield: 11.0 g (35.3 mmol, 93 %)
TLC: R_f = 0.18 (CH₂Cl₂/MeOH: 9/1)

¹H-NMR: (400 MHz, DMSO-d_6)
  δ [ppm] = 9.99 (br s, 1H, NH₂ (exchange)), 8.86 (br s, 1H, NH₂ (exchange)), 7.98 (d, 1H, H6, J = 7.96), 6.23 (d, 1H, H5, J = 7.96), 6.08 (ψt, 1H, H1', J = 6.59), 5.18 (m, 1H, H3'), 4.30 – 4.21 (m, 3H, H4', H5'), 2.43 (m, 2H, H2'), 2.06 (s, 3H, acetyl-CH₃), 2.03 (s, 3H, acetyl-CH₃).

¹³C-NMR: (100 MHz, DMSO-d_6)
  δ [ppm] = 170.1 (acetyl-C=O), 170.0 (acetyl-C=O), 159.6 (C4), 146.9 (C2), 144.3 (C6), 94.3 (C5), 86.3 (C1'), 82 (C4'), 73.7 (C3'), 63.5 (C5'), 36.6 (C2'), 20.7 (acetyl-CH₃), 20.6 (acetyl-CH₃).

ESI(+)-MS: m/z   311.9 [M + H⁺]

5-Iodo-(3',5'-O-diacetyl)-2'-deoxycytidine

65
$C_{13}H_{16}IN_3O_6$
437.19 g/mol

3.5 g (10 mmol) of nucleoside **64** were dissolved in 35 ml tetrachlorocarbon and 35 ml glacial acetic acid, then treated with 1.52 g (6 mmol) iodine and the dark red solution was heated up to 40 °C with stirring. A freshly prepared solution of 1.58 g (9 mmol) iodic acid in 4 ml water was added and the mixture stirred at 40 °C for about 4 h while it turned cloudy yellow. After solvent removal, the obtained residue was dissolved in 200 ml CH_2Cl_2 and washed with sat. $NaHCO_3$-solution and water until the filtrate reached pH 7. The excess iodine was removed by washing the organic layer with 5 % aq. $NaHSO_3$-solution. After drying the organic layer over $MgSO_4$, it was filtered and concentrated. The yellow crystalline product was obtained after column chromatography with CH_2Cl_2/MeOH (3 - 10 % MeOH).

Yield: 3.46 g (7.9 mmol, 79 %)
TLC: R_f = 0.31 (CH_2Cl_2/MeOH: 9/1)

^{1}H-NMR: (250 MHz, DMSO-d_6)
δ [ppm] = 7.94 (s, 1H, H6), 7.91 (br s, 1H, NH_2(exchange)), 6.72 (br s, 1H, NH_2(exchange)), 6.11 (ψt, 1H, H1'), 5.17 (m, 1H, H3'), 4.29 - 4.11 (m, 3H, H4', H5'), 2.33 (m, 2H, H2'), 2.09 (s, 3H, acetyl-CH_3), 2.06 (s, 3H, acetyl-CH_3).

^{13}C-NMR: (63 MHz, DMSO-d_6)
δ [ppm] = 170.1 (acetyl-C=O), 170.0 (acetyl-C=O), 163.7 (C4), 153.6 (C2), 146.8 (C6), 85.6 (C1'), 81.5 (C4'), 74.2 (C3'), 63.7 (C5'), 57.3 (C5), 36.8 (C2'), 20.7 (acetyl-CH_3), 20.7 (acetyl-CH_3).

ESI(+)-MS: m/z 437.9 [M + H$^+$]

5-Iodo-2'-deoxycytidine

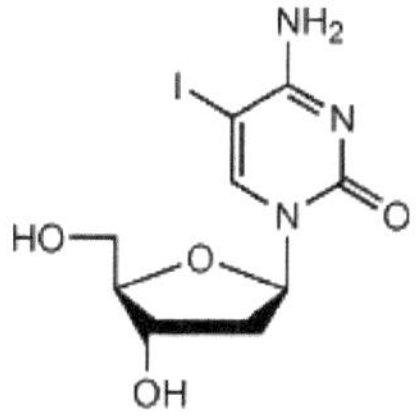

66
$C_9H_{12}IN_3O_4$
353.11 g/mol

4.17 g (9.5 mmol) of compound **65** were dissolved in 100 ml of a freshly prepared 0.5 M sodium methoxide/MeOH solution and stirred at room temperature for 1 h. The mixture was neutralized by the addition of Dowex (H^+-form) and concentrated. The yellow crystalline crude product was purified on a silica gel column using CH_2Cl_2/MeOH 8/2 as eluent.

Yield: 3.12 g (8.8 mmol, 93 %)
TLC: R_f = 0.17 (CH_2Cl_2/MeOH: 8/2)

^{1}H-NMR: (400 MHz, DMSO-d_6)
δ [ppm] = 8.28 (s, 1H, H6), 7.77 (br s, 1H, NH_2(exchange)), 6.58 (br s, 1H, NH_2(exchange)), 6.07 (ψt, 1H, H1'), 5.21 (d, 1H, 3'-OH), 5.11 (t, 1H, 5'-OH), 4.20 (m, 1H, H3'), 3.77 (m, 1H, H4'), 3.55 (m, 2H, H5'), 2.19 - 1.92 (m, 2H, H2').

^{13}C-NMR: (100 MHz, DMSO-d_6)
δ [ppm] = 164 (C4), 154.2 (C2), 147.8 (C6), 87.1 (C4'), 85.1 (C1'), 69.7 (C3'), 60.6 (C5'), 56.5 (C5), 40.5 (C2').

ESI(+)-MS: *m/z* 353.8 [M + H$^+$]

5-Iodo-4N-(N,N-dimethylaminomethylidenyl)-2'-deoxycytidine

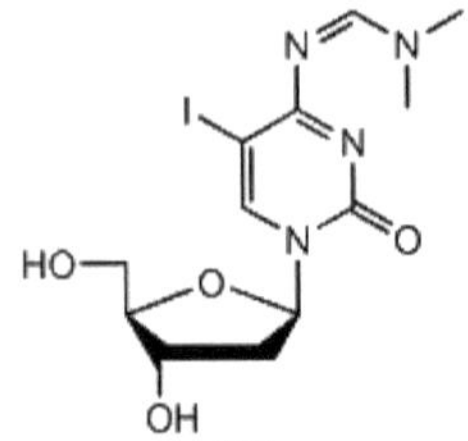

67
$C_{12}H_{17}IN_4O_4$
408.19 g/mol

2.0 g (5.66 mmol) of nucleoside **66** were dissolved in 20 ml dry DMF. Under argon, 15 ml (113 mmol) of N,N-dimethylformamide dimethylacetal were added and the reaction mixture was heated up to 55 °C for 2.5 h. The solvent was evaporated and the residue purified on a silica gel column using CH_2Cl_2/MeOH (5 – 10 % MeOH) as eluent to give the product as yellowish crystals.

Yield: 1.43 g (3.5 mmol, 62 %)
TLC: R_f = 0.1 (CH_2Cl_2/MeOH: 9/1)

^{1}H-NMR: (400 MHz, DMSO-d_6)
δ [ppm] = 8.56 (s, 1H, formamidino-CH), 8.46 (s, 1H, H6), 6.09 (t, 1H, H1', J = 6.32), 5.22 (d, 1H, H3'), 5.12 (t, 1H, H5'), 4.23 (m, 1H, H3'), 3.61 (m, 2H, H5'), 3.20 (s, 3H, CH$_3$), 3.13 (s, 3H, CH$_3$), 2.24 – 1.96 (m, 2H, H2').

^{13}C-NMR: (100 MHz, DMSO-d_6)
δ [ppm] = 168 (C4), 158.3 (formamidino-C=N), 154.2 (C2), 147.3 (C6), 87.5 (C4'), 85.7 (C1'), 69.8 (C3'), 68.7 (C5), 60.7 (C5'), 41 (formamidino-CH$_3$), 40.9 (C2'), 34.9 (formamidino-CH$_3$).

ESI(+)-MS: m/z 409.1 [M + H$^+$]

5-Iodo-4N-(N,N-dimethylaminomethylidenyl)-5'-O-benzoyl-2'-deoxycytidine

68
$C_{19}H_{21}IN_4O_5$
512.30 g/mol

1.0 g (2.45 mmol) of well-dried nucleoside **67** was dissolved in 45 ml dry pyridine and 10 ml dry DMF. The pale yellow solution was cooled down to 0 °C, then 385 µl (3.28 mmol) benzoyl chloride in 4 ml dry pyridine were added dropwise via syringe during 2 h. The reaction was allowed to proceed for 30 min at room temperature, then the solvents were evaporated and the residual oil purified on column using CH_2Cl_2/MeOH (5 - 10 % MeOH) furnishing the pure crystalline product.

Yield: 750 mg (1.46 mmol, 60 %)
TLC: R_f = 0.13 (CH_2Cl_2/MeOH: 95/5)

^{1}H-NMR: (400 MHz, DMSO-d_6)
δ [ppm] = 8.56 (s, 1H, formamidino-CH), 8.07 (s, 1H, H6), 8.01 (d, 2H, benzoyl-H_A, J = 7.33), 7.68 (t, 1H, benzoyl-H_C, J = 7.33), 7.55 (t, 2H, benzoyl-H_B, J = 7.58), 6.14 (ψt, 1H, H1', J = 6.57), 5.45 (d, 1H, 3'-OH, J = 4.29), 4.51 (m, 2H, H5'), 4.35 (m, 1H, H3'), 4.16 (m, 1H, H4'), 3.21 (s, 3H, CH_3), 3.12 (s, 3H, CH_3), 2.33 – 2.12 (m, 2H, H2').

^{13}C-NMR: (100 MHz, DMSO-d_6)
δ [ppm] = 168.1 (C4), 165.6 (benzoyl-C=O), 158.3 (formamidino-C=N), 154 (C2), 146.5 (C6), 133.5 (benzoyl-C_C), 129.2 (benzoyl-C_A), 128.9 (benzoyl-C_B), 86.1 (C1'), 84.4 (C4'), 70.5 (C3'), 69.1 (C5), 64.5 (C5'), 40.9 (formamidino-CH_3), 40.1 (C2'), 34.9 (formamidino-CH_3).

ESI(+)-MS: m/z   513.1 [M + H$^+$]

5-Iodo-4N-(N,N-dimethylaminomethylidenyl)-5'-O-benzoyl-3'-O-(2-cyanoethyl)-2'-deoxycytidine

69
$C_{22}H_{24}IN_5O_5$
565.36 g/mol

In a well dried sealable Erlenmeyer flask with magnetic triangle stirrer bar, 1.02 g (2 mmol) of nucleoside **68** were dissolved under argon in 2.64 ml (40 mmol) freshly distilled acrylonitrile, 12 ml *tert*-butanol and 6 ml abs. DMF. After 2 min of stirring, 652 mg (2 mmol) of cesium carbonate were added and the pale yellow suspension was vigorously agitated for about 3 h at room temperature. The mixture was taken up in 100 ml methylene chloride then and filtered over Celite. The filtrate was concentrated and the residue purified on a silica gel column using CH_2Cl_2/MeOH (2 – 10 % MeOH) to give the product as yellow foam.

Yield: 1.0 g (1.77 mmol, 88 %)
TLC: R_f = 0.26 (CH_2Cl_2/MeOH: 95/5)

^{1}H-NMR: (400 MHz, $CDCl_3$)
δ [ppm] = 8.68 (s, 1H, formamidino-NH), 8.06 (s, 1H, H6), 8.01 (m, 2H, benzoyl-H_A), 7.56 (ψt, 1H, benzoyl-H_C, J = 7.33), 7.44 (ψt, 2H, benzoyl-H_B, J = 7.57), 6.21 (dd, 1H, H1', J = 6.06 and J = 7.32), 4.59 (m, 2H, H5'), 4.42 (m, 1H, H3'), 4.18 (m, 1H, H4'), 3.72 (t, 2H, O-C$\underline{H}_2$, J = 6.06), 3.18 (s, 3H, CH_3), 3.16 (s, 3H, CH_3), 2.61 (t, 2H, C$\underline{H}_2$CN, J = 6.06), 2.77 – 2.59 (m, 2H, H2').

^{13}C-NMR: (100 MHz, $CDCl_3$)
δ [ppm] = 168.9 (C4), 166.2 (benzoyl-C=O), 158.9 (formamidino-C=N), 155.3 (C2), 146 (C6), 133.6 (benzoyl-C_C), 129.7 (benzoyl-C_{quart}), 129.2 (benzoyl-C_A), 128.8 (benzoyl-C_B), 117.5 (CN), 87.2 (C1'), 83 (C4'), 80.3 (C3'), 69.4 (C5), 64.4 (C5'), 64.3 (-O-$\underline{C}H_2$), 41.5 (formamidino-CH_3), 38.8 (C2'), 35.6 (formamidino-CH_3), 19 ($\underline{C}H_2$CN).

ESI(+)-MS: m/z 566.1 [M + H$^+$]

5-Iodo-3'-O-(2-cyanoethyl)-2'-deoxycytidine

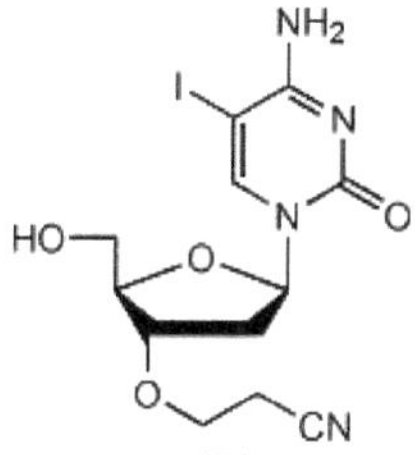

70
$C_{12}H_{15}IN_4O_4$
406.18 g/mol

250 mg (0.44 mmol) of nucleoside **69** were dissolved in 50 ml sat. methanolic ammonia and stirred in a sealable vessel at room temperature for 2.5 h until complete consumption of the starting material. Concentration of the mixture gave the crude product which was purified on a silica gel column using CH_2Cl_2/MeOH 95/5 as eluent.

Yield: 147 mg (0.36 mmol, 82 %)
TLC: R_f = 0.16 (CH_2Cl_2/MeOH: 9/1)

^{1}H-NMR: (400 MHz, DMSO-d_6)
δ [ppm] = 8.24 (s, 1H, H6), 7.82 (br s, 1H, NH(exchange)), 6.62 (br s, 1H, NH(exchange)), 6.05 (dd, 1H, H1', J = 6.06 and J = 7.32), 5.18 (t, 1H, 5'-OH, J = 5.05), 4.12 (m, 1H, H3'), 3.94 (m, 1H, H4'), 3.62 (t, 2H, -O-C$\underline{H}_2$, J = 6.06) , 3.59 (m, 2H, H5'), 2.77 (t, 2H, C$\underline{H}_2$CN, J = 6.06), 2.3 – 2.06 (m, 2H, H2').

^{13}C-NMR: (100 MHz, DMSO-d_6)
δ [ppm] = 163.7 (C4), 153.8 (C2), 147.2 (C6), 119.2 (CN), 85.3 (C1'), 84.7 (C4'), 79.1 (C3'), 63.6 (-O-$\underline{C}$H$_2$), 61.1 (C5'), 56.8 (C5), 37.6 (C2'), 18.3 ($\underline{C}$H$_2$CN).

ESI(-)-MS: m/z 405.1 [M - H$^+$]

Elemental analysis: calculated: C, 35.48; H, 3.72; N, 13.79; O, 15.76
found: C, 35.35; H, 3.93; N, 13.62

5-[3-Trifluoroacetamido-prop-1-ynyl]-3'-O-(2-cyanoethyl)-2'-deoxycytidine

71
$C_{17}H_{18}F_3N_5O_5$
429.35 g/mol

Under argon, 580 mg (1.43 mmol) of compound **70** were dissolved in 12 ml abs. DMF and 992 µl (7.14 mmol) TEA. The mixture was degassed in vacuum three times, then 54 mg (0.29 mmol) copper iodide and 165 mg (0.14 mmol) tetrakis-(triphenylphosphin)-palladium were added. At last, 502 µl (2.86 mmol) N-propargyltrifluoroacetamide was injected via syringe through the septum and the mixture stirred in the dark for 24 h under argon. After concentration of the crude orange-brownish mixture, the residue was diluted in 100 ml CH_2Cl_2 and washed once with 100 ml of a 5 % aq. disodium EDTA-solution. The aqueous layer was extracted three times with 50 ml 2-butanone for quantitative isolation of the product. The combined organic layers were dried over $MgSO_4$, filtered and concentrated again. The residue was purified on a short silica gel column using CH_2Cl_2/MeOH (0 - 10 % MeOH) as eluent, yielding the product as yellow glassy oil.

Yield: 522 mg (1.21 mmol, 85 %)
TLC: R_f = 0.32 (CH_2Cl_2/MeOH: 85/15)

^{1}H-NMR: (400 MHz, DMSO-d_6)
δ [ppm] = 9.96 (br t, 1H, TFA-N$\underline{H}$, J = 5.05), 8.13 (s, 1H, H6), 7.85 (br s, 1H, NH_2(exchange)), 6.88 (br s, 1H, NH_2(exchange)), 6.08 (ψt, 1H, H1', J = 6.06), 5.17 (t, 1H, 5'-OH, J = 4.55), 4.28 (d, 2H, propargyl-CH_2, J = 5.31), 4.11 (m, 1H, H3'), 3.95 (m, 1H, H4'), 3.62 (t, 2H, -O-C$\underline{H}_2$, J = 6.06), 3.58 (m, 2H, H5'), 2.76 (t, 2H, C$\underline{H}_2$CN, J = 6.06), 2.33 - 2.03 (m, 2H, H2').

^{13}C-NMR: (100 MHz, DMSO-d_6)
δ [ppm] = 164.4 (C4), 156 (TFA-$\underline{C}$=O), 153.4 (C2), 144.4 (C6), 119.2 (CN), 117.3

($\underline{C}F_3$), 104.1 (C5), 90.1 (propargyl-$\underline{C}$-CH$_2$), 85.5 (C1'), 84.9 (C4'), 79.2 (C3'), 74.5 (propargyl-$\underline{C}$-C5), 63.6 (-O-$\underline{C}H_2$), 61.3 (C5'), 37.7 (C2'), 30 (propargyl-$\underline{C}H_2$), 18.3 ($\underline{C}H_2$CN).

ESI(+)-MS: m/z   430.0 [M + H$^+$]

5-[3-Amino-prop-1-ynyl]-3'-O-(2-cyanoethyl)-2'-deoxycytidine

72
C$_{15}$H$_{19}$N$_5$O$_4$
333.34 g/mol

520 mg (1.2 mmol) of compound **71** were dissolved in 20 ml MeOH and 10 ml 25 % NH$_4$OH. The mixture was stirred at room temperature for 5 h, concentrated and the residue purified on a short silica gel column with CH$_2$Cl$_2$/MeOH (10 – 100 % MeOH) to give the product as yellow glassy oil.

Yield: 343 mg (1.03 mmol, 86 %)
TLC: R$_f$ = 0.05 (CH$_2$Cl$_2$/MeOH: 9/1)

^{1}H-NMR: (400 MHz, DMSO-d_6)
δ [ppm] = 8.07 (s, 1H, H6), 7.76 (br s, 1H, NH$_2$(exchange)), 6.94 (br s, 1H, NH$_2$(exchange)), 6.09 (dd, 1H, H1', J = 5.81 and J = 7.33), 5.25 (br s, 1H, 5'-OH), 4.12 (m, 1H, H3'), 3.96 (m, 1H, H4'), 3.62 (t, 2H, -O-C$\underline{H}_2$, J = 6.05), 3.60 (m, 2H, H5'), 3.42 (br s, 2H, propargyl-C$\underline{H}_2$-NH$_2$), 2.77 (t, 2H, C$\underline{H}_2$CN, J = 6.05), 2.31 – 2.02 (m, 2H, H2').

^{13}C-NMR: (100 MHz, DMSO-d_6)
δ [ppm] = 164.4 (C4), 153.5 (C2), 143.6 (C6), 119.3 (CN), 105.2 (C5), 90.2 (propargyl-$\underline{C}$-CH$_2$), 85.4 (C1'), 84.9 (C4'), 79.4 (C3'), 76.7 (propargyl-$\underline{C}$-C5), 63.7 (-O-$\underline{C}H_2$), 61.3 (C5'), 37.7 (C2'), 30.1 (propargyl-$\underline{C}H_2$), 18.3 ($\underline{C}H_2$CN).

ESI(+)-MS: m/z   334.1 [M + H$^+$]

5-Iodo-5'-O-benzoyl-2'-deoxyuridine

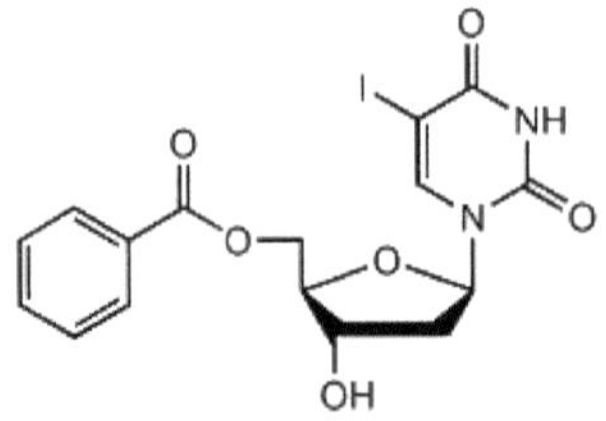

74
$C_{16}H_{15}IN_2O_6$
458.20 g/mol

7.72 g (21.8 mmol) of well-dried nucleoside **73** were dissolved in 30 ml abs. pyridine and cooled down to -20 °C. Under argon, 2.66 ml (23 mmol) of benzoyl chloride, diluted in 7.5 ml dry CH_2Cl_2, were added dropwise within 10 min. Conversion of the starting material was complete after stirring the mixture for 2 h at -20 °C and addition of 3 ml MeOH. The solvent was removed and the resulting residue purified on column chromatography using CH_2Cl_2/MeOH (0 - 10 % MeOH) to give the pure product as colorless foam.

Yield: 5.06 g (11 mmol, 51 %)
TLC: R_f = 0.5 (CH_2Cl_2/MeOH: 9/1)

^{1}H-NMR: (250 MHz, DMSO-d_6)
δ [ppm] = 11.67 (br s, 1H, N3-H), 8.01 (s, 1H, H6), 7.98 - 7.52 (m, 5H, benzoyl-H), 6.13 (ψt, 1H, H1'), 5.48 (br s, 1H, 3'-OH), 4.49 (m, 2H, H5'), 4.36 (m, 1H, H3'), 4.1 (m, 1H, H4'), 2.35 - 2.14 (m, 2H, H2').

^{13}C-NMR: (63 MHz, DMSO-d_6)
δ [ppm] = 165.6 (benzoyl-C=O), 160.4 (C4), 150.1 (C2), 144.5 (C6), 133.5 (benzoyl-C_C), 129.3 (benzoyl-C_A), 128.9 (benzoyl-C_B), 85.1 (C4'), 84.1 (C1'), 70.3 (C5'), 69.8 (C3'), 64.4 (C5), 40 (C2').

ESI(-)-MS: m/z 457.1 [M - H$^+$]

5-Iodo-2'-deoxy-(5'-O,3N-dibenzoyl)uridine

76
$C_{23}H_{19}IN_2O_7$
562.31 g/mol

2.5 g (5.46 mmol) of the compound **74** were coevaporated three times with abs. pyridine and dried overnight in vacuum. The nucleoside was dissolved in 50 ml dry pyridine and treated with 4.7 ml (27.3 mmol) N-ethyldiisopropylamine and 1.03 ml (8.2 mmol) chlorotrimethylsilane. After stirring the solution at room temperature for 30 min, 0.95 ml (8.2 mmol) benzoyl chloride were added and stirring continued for 1 h. The reaction was stopped by the addition of 3 ml MeOH with subsequent solvent removal. The oily residue was taken up in 100 ml CH_2Cl_2, washed with each 100 ml of water, 5 % aq. $NaHCO_3$-solution and brine. The organic layer was dried over $MgSO_4$, filtered and the solvent evaporated. The residue was dissolved again in 50 ml CH_2Cl_2/MeOH 1/1 and treated with 0.5 ml of trifluoroacetic acid while stirring the mixture at room temperature for 30 min. The mixture was diluted in 100 ml CH_2Cl_2 then and washed twice with 100 ml water each. The aqueous layer was back-extracted with CH_2Cl_2 (4 x 50 ml) and the combined organic layers were dried over $MgSO_4$, filtered and concentrated. The crude product was purified on column chromatography using CH_2Cl_2/MeOH (2 - 5 % MeOH) to furnish the pure compound as yellowish foam.

Yield: 2.87 g (5.1 mmol, 93 %)
TLC: R_f = 0.22 (CH_2Cl_2/MeOH: 98/2)

^{1}H-NMR: (400 MHz, DMSO-d_6)
δ [ppm] = 8.20 (s, 1H, H6), 8.04 - 7.54 (m, 10H, benzoyl-H), 6.11 (ψt, 1H, H1'), 5.51 (d, 1H, 3'-OH), 4.52 (m, 2H, H5'), 4.39 (m, 1H, H3'), 4.16 (m, 1H, H4'), 2.46 - 2.24 (m, 2H, H2').

^{13}C-NMR: (100 MHz, DMSO-d_6)
δ [ppm] = 168.7 (benzoyl-C=O), 165.6 (benzoyl-C=O), 159.2 (C2), 148.7 (C4),

145.3 (C6), 135.7 (benzoyl-$C_{C'}$), 133.6 (benzoyl-C_C), 130.7 (benzoyl-$C_{A'}$), 129.5 (benzoyl-C_A), 129.3 (benzoyl-$C_{B'}$), 128.9 (benzoyl-C_B), 86.1 (C1'), 84.5 (C4'), 70.1 (C3'), 68.6 (C5), 64.3 (C5'), 39.7 (C2').

ESI(+)-MS: *m/z* 563.0 [M + H$^+$]

5-Iodo-3'-O-(2-cyanoethyl)-2'-deoxy-(5'-O,3N-dibenzoyl)uridine

77
$C_{26}H_{22}IN_3O_7$
615.37 g/mol

In a well dried sealable Erlenmeyer flask with magnetic stirrer and triangle stirring bar, 2.0 g (3.56 mmol) of nucleoside **76** were dissolved under argon in 4.6 ml (71.2 mmol) freshly distilled acrylonitrile and 15 ml *tert*-butanol. The solution was treated with 1.16 g (3.56 mmol) of cesium carbonate and vigorously agitated for about 2 h at room temperature. The mixture was taken up in 100 ml CH_2Cl_2 and the insoluble material removed via filtration through Celite. The filtrate was concentrated furnishing the crude product which was purified on a silica gel column using CH_2Cl_2/MeOH (0 – 1 % MeOH) as eluent.

Yield: 1.0 g (1.63 mmol, 46 %)
TLC: R_f = 0.67 (CH_2Cl_2/MeOH: 98/2)

^{1}H-NMR: (400 MHz, DMSO-d_6)
δ [ppm] = 8.22 (s, 1H, H6), 8.02 (ψd, 4H, benzoyl-$H_{A,A'}$, *J* = 7.32), 7.79 (t, 1H, benzoyl-H_C, *J* = 7.32), 7.7 (t, 1H, benzoyl-$H_{C'}$), 7.58 (m, 4H, benzoyl-$H_{B,B'}$, *J* = 7.68, *J* = 8.05), 6.08 (ψt, 1H, H1', *J* = 6.59), 4.54 (m, 2H, H5', *J* = 3.66), 4.35 (m, 2H, H3', H4'), 3.69 (t, 2H, -O-C$\underline{H}_2$, *J* = 6.22), 2.79 (t, 2H, C$\underline{H}_2$CN, *J* = 6.22), 2.46 (m, 2H, H2').

^{13}C-NMR: (100 MHz, DMSO-d_6)

δ [ppm] = 166.5 (benzoyl-C=O), 166.3 (benzoyl-C=O), 158.9 (C2), 148.9 (C4), 145.3 (C6), 135.7 (benzoyl-$C_{C'}$), 133.7 (benzoyl-C_C), 130.7 (benzoyl-$C_{A'}$), 129.5 (benzoyl-C_A), 129.4 (benzoyl-$C_{B'}$), 128.9 (benzoyl-C_B), 119.2 (CN), 86.2 (C1'), 82 (C4'), 78.7 (C3'), 68.8 (C5), 64.5 (C5'), 63.9 (O-$\underline{C}H_2$), 38.4 (C2'), 18.2 ($\underline{C}H_2CN$).

ESI(+)-MS: m/z   616.3 [M + H$^+$]

5-Iodo-3'-O-(2-cyanoethyl)-2'-deoxyuridine

78
$C_{12}H_{14}IN_3O_5$
407.16 g/mol

1.0 g (1.63 mmol) of nucleoside **77** was dissolved in 20 ml MeOH and 25 ml 32 % aqueous ammonia. The mixture was stirred in a sealable vessel for 24 h at room temperature, then the solvents were evaporated, the residue lyophyllized and purified on column chromatography using CH_2Cl_2/MeOH 95/5 to give the pure compound as colorless foam.

Yield: 581 mg (1.43 mmol, 88 %)
TLC: R_f = 0.46 (CH_2Cl_2/MeOH: 85/15)

^{1}H-NMR: (400 MHz, DMSO-d_6)
δ [ppm] = 11.7 (br s, 1H, H3), 8.37 (s, 1H, H6), 6.07 (ψt, 1H, H1'), 5.24 (t, 1H, 5'-OH, J = 4.8), 4.15 (m, 1H, H3'), 3.97 (m, 1H, H4'), 3.63 (t, 2H, -O-$\underline{C}H_2$, J = 6.09), 3.61 (m, 2H, H5'), 2.78 (t, 2H, $\underline{C}H_2CN$, J = 6.09), 2.24 (m, 2H, H2').

^{13}C-NMR: (400 MHz, DMSO-d_6)
δ [ppm] = 160.8 (C4), 150.4 (C2), 145.1 (C6), 119.4 (CN), 85.3 (C4'), 84.8 (C1'), 79.3 (C3'), 69.9 (C5), 63.7 (-O-$\underline{C}H_2$), 61.3 (C5'), 37.2 (C2'), 18.4 ($\underline{C}H_2CN$).

ESI(-)-MS: m/z   406.0 [M - H$^+$]

5-[3-Trifluoroacetamido-prop-1-ynyl]-3'-O-(2-cyanoethyl)-2'-deoxyuridine

79
$C_{17}H_{17}F_3N_4O_6$
430.34 g/mol

Under argon, 302 mg (0.74 mmol) of the starting material **78** were dissolved in 10 ml abs. DMF and 514 µl (3.7 mmol) TEA. The mixture was degassed under vacuum three times, then 28 mg (0.15 mmol) copper iodide and 86 mg (0.074 mmol) tetrakis-(triphenylphosphin)palladium were added. After stirring the mixture for 2 min, 260 µl (1.84 mmol) of N-propargyltrifluoroacetamide were injected via syringe through the septum and stirring continued for 24 h under argon and in the dark. The crude orange-brownish mixture was concentrated under reduced pressure then, diluted in 100 ml methylene chloride and washed once with 100 ml of a 5 % aqueous disodium EDTA-solution. The aqueous layer was extracted three times with 50 ml 2-butanone for quantitative product isolation. The combined organic layers were dried over $MgSO_4$, filtered and concentrated again. The oily residue was purified on a short silica gel column using CH_2Cl_2/MeOH (0 – 10 % MeOH) as eluent to give the product as yellow crystals.

Yield: 290 mg (0.67 mmol, 91 %)
TLC: R_f = 0.41 (CH_2Cl_2/MeOH: 85/15)

^{1}H-NMR: (400 MHz, DMSO-d_6)
δ [ppm] = 11.67 (s, 1H, H3), 10.06 (t, 1H, TFA-N<u>H</u>, J = 5.56), 8.18 (s, 1H, H6), 6.06 (ψt, 1H, H1'), 5.18 (t, 1H, 5'-OH, J = 5.05), 4.32 (d, 2H, propargyl-C<u>H</u>$_2$, J = 5.30), 4.14 (m, 1H, H3'), 3.95 (m, 1H, H4'), 3.63 (t, 2H, -O-C<u>H</u>$_2$, J = 6.06), 3.59 (m, 2H, H5'), 2.77 (t, 2H, C<u>H</u>$_2$CN, J = 6.06), 2.25 (m, 2H, H2').

^{13}C-NMR: (100 MHz, DMSO-d_6)
δ [ppm] = 161.5 (C4), 155.9 (TFA-<u>C</u>=O), 149.4 (C2), 144.1 (C6), 119.2 (CN), 117

($\underline{C}$F$_3$), 97.8 (C5), 87.6 (propargyl-$\underline{C}$-CH$_2$-NH), 84.9 (C4'), 84.7 (C1'), 79.1 (C3'), 75.3 (propargyl-$\underline{C}$-C5), 63.6 (-O-$\underline{C}$H$_2$), 61.1 (C5'), 37 (C2'), 29.5 (propargyl-$\underline{C}$H$_2$-NH), 18.3 ($\underline{C}$H$_2$CN).

ESI(+)-MS: m/z   431.1 [M + H$^+$]

5-[3-Amino-prop-1-ynyl]-3'-O-(2-cyanoethyl)-2'-deoxyuridine

80
C$_{15}$H$_{18}$N$_4$O$_5$
334.33 g/mol

294 mg (0.68 mmol) of nucleoside **79** were dissolved in 15 ml MeOH and 10 ml of 32 % aq. ammonia. The pale yellow solution was stirred at room temperature for 4 h until completion of the reaction. The mixture was concentrated and the resulting residue purified on a short silica gel column using CH$_2$Cl$_2$/MeOH (10 – 100 % MeOH) as eluent to give the product as yellow glassy oil.

Yield: 228 mg (0.68 mmol, 100 %)
TLC: R$_f$ = 0.12 (CH$_2$Cl$_2$/MeOH: 9/1)

^{1}H-NMR: (400 MHz, DMSO-d_6)
δ [ppm] = 9.19 (s, 1H, H3), 8.24 (s, 1H, H6), 6.06 (ψt, 1H, H1'), 4.14 (m, 1H, H3'), 3.98 (m, 1H, H4'), 3.94 (d, 2H, propargyl-C$\underline{H}_2$, J = 5.30), 3.60 (t, 2H, -O-C$\underline{H}_2$, J = 6.06), 3.59 (m, 2H, H5'), 2.76 (t, 2H, C$\underline{H}_2$CN, J = 5.86), 2.31 – 2.15 (m, 2H, H2').

^{13}C-NMR: (100 MHz, DMSO-d_6)
δ [ppm] = 161.6 (C4), 149.5 (C2), 144.9 (C6), 119.3 (CN), 97.3 (C5), 87.6 (propargyl-$\underline{C}$-CH$_2$), 85.2 (C1'), 85.1 (C4'), 79.3 (C3'), 78.9 (propargyl-$\underline{C}$-C5), 63.7 (-O-$\underline{C}$H$_2$), 61.3 (C5'), 37.3 (C2'), 29.3 (propargyl-$\underline{C}$H$_2$), 18.4 ($\underline{C}$H$_2$CN).

ESI(+)-MS: m/z   335.1 [M + H$^+$]

5'-O-Benzoyl-2'-deoxythymidine

82
$C_{17}H_{18}N_2O_6$
346.33 g/mol

9.69 g (40 mmol) of 2'-deoxythymidine were dried by coevaporation with pyridine and in vacuum overnight. The nucleoside was dissolved in 130 ml abs. pyridine under argon and cooled down to -20 °C then. 5.16 ml (44 mmol) of benzoyl chloride diluted in 20 ml abs. pyridine were added dropwise via syringe within 2 h. The reaction was allowed to proceed at -20 °C for 1 h and then overnight at room temperature. After addition of 5 ml MeOH, the mixture was concentrated and coevaporated twice with toluene. The residue was purified via column chromatography with CH_2Cl_2/MeOH (5 – 10 % MeOH) to give the product as colorless crystals.

Yield: 11.51 g (33.23 mmol, 83 %)
TLC: R_f = 0.31 (CH_2Cl_2/MeOH: 9/1)

^{1}H-NMR: (400 MHz, DMSO-d_6)
δ [ppm] = 11.3 (br s, 1H, H3), 7.98 (ψd, 2H, benzoyl-H_A, J = 8.42), 7.68 (ψt, 1H, benzoyl-H_C, J = 7.32), 7.54 (ψt, 2H, benzoyl-H_B, J = 7.32), 7.39 (s, 1H, H6), 6.21 (ψt, 1H, H1', J = 6.95), 4.68 (d, 2H, J = 11.87, O-CH_2-S), 4.57 – 4.39 (m, 2H, H5'), 4.4 (m, 1H, H3'), 4.05 (m, 1H, H4'), 2.28 – 2.14 (m, 2H, H2'), 1.59 (s, 3H, CH_3).

^{13}C-NMR: (100 MHz, DMSO-d_6)
δ [ppm] = 165.6 (benzoyl-C=O), 163.6 (C4), 150.4 (C2), 135.6 (C6), 133.5 (benzoyl-C_C), 129.4 (benzoyl-C_A), 129.2 (benzoyl-C_{quart}), 128.9 (benzoyl-C_B), 109.7 (C5), 83.8 (C4'), 83.6 (C1'), 70.2 (C3'), 64.4 (C5'), 38.8 (C2'), 11.8 (CH_3).

ESI(+)-MS: m/z 347.1 [M + H$^+$]

5'-O-Benzoyl-2'-deoxy-3'-O-(methylthiomethyl)thymidine

83
$C_{19}H_{22}N_2O_6S$
406.45 g/mol

11.0 g (31.8 mmol) of compound **82** were dissolved in 106 ml DMSO, 21.3 ml glacial acetic acid and 69 ml acetic anhydride and stirred under argon at room temperature for 20 h. The solvent was removed under reduced pressure then and the orange residue diluted in 250 ml CH_2Cl_2, washed twice with each 200 ml of water and sat. aqueous $NaHCO_3$. The organic layer was dried over $MgSO_4$, filtered and concentrated in vacuum. The resulting yellowish foam was purified on column chromatography using CH_2Cl_2/MeOH (0 - 2 % MeOH) as eluent. The solvent residues were removed by recrystallization in toluene to give the pure product as colorless crystals.

Yield: 7.95 g (19.6 mmol, 62 %)
TLC: R_f = 0.66 (CH_2Cl_2/MeOH: 9/1)

^{1}H-NMR: (400 MHz, $CDCl_3$)
δ [ppm] = 8.65 (br s, 1H, H3), 8.03 (d, 2H, benzoyl-H_A, J = 7.82), 7.61 (t, 1H, benzoyl-H_C, J = 7.36), 7.47 (t, 2H, benzoyl-H_B, J = 7.72), 7.23 (s, 1H, H6), 6.29 (dd, 1H, H1', J = 5.97 and J = 7.77), 4.68 (d, 2H, J = 11.87, O-C$\underline{H}_2$-S), 4.61 (m, 2H, H5') 4.57 (m, 1H, H3'), 4.36 (q, 1H, H4'. J = 3.54), 2.51 (m, 1H, H2'), 2.15 (s, 3H, SCH_3), 2.13 (m, 1H, H2'), 1.67 (s, 3H, CH_3).

^{13}C-NMR: (100 MHz, $CDCl_3$)
δ [ppm] = 166.2 (benzoyl-C=O), 163.65 (C4), 150.2 (C2), 134.9 (C6), 133.8 (benzoyl-C_C), 129.7 (benzoyl-C_A), 129.5 (benzoyl-C_{quart}), 128.9 (benzoyl-C_B), 111.5 (C5), 85.4 (C1'), 82.4 (C4'), 76.0 (C3'), 74.1 (-O-$\underline{C}H_2$-S-), 64.3 (C5'), 37.9 (C2'), 14.0 (SCH_3) 12.4 (CH_3).

ESI(+)-MS: m/z 406.9 [M + H$^+$]

5'-O-Benzoyl-3'-O-[(2-cyanoethoxy)methyl]-2'-deoxythymidine

85
$C_{21}H_{23}N_3O_7$
429.42 g/mol

After drying 2.03 g (5 mmol) of nucleoside **83** in vacuum overnight, it was dissolved under argon in 25 ml dry 1,2-dichloroethane followed by addition of preactivated molecular sieves (3 Å). The mixture was treated with 696 µl (5 mmol) triethylamine and stirred for 2 h at room temperature. The suspension was cooled down to 0 °C and 456 µl (5.5 mmol) sulfuryl chloride, diluted in 2 ml dry 1,2-dichloroethane were added dropwise within 5 min. The reaction was allowed to proceed for 1 h at 0 °C, then 715 µl (11 mmol) 3-hydroxy-propionitrile were added via syringe within 10 min. The mixture was warmed up to room temperature then and stirred overnight under argon. The clear-yellow suspension was diluted with 100 ml methylene chloride, washed once with 100 ml sat. aqueous $NaHCO_3$, dried over Na_2SO_4, filtered and concentrated. The residue of the organic layer was taken up in 100 ml CH_2Cl_2 again, filtered over Celite and concentrated under reduced pressure. The resulting syrup was purified on a silica gel column using CH_2Cl_2/MeOH (2 – 5 % MeOH) to give the product as colorless foam.

Yield: 1.49 g (3.47 mmol, 69 %)
TLC: R_f = 0.46 (CH_2Cl_2/MeOH: 9/1)

^{1}H-NMR: (400 MHz, $CDCl_3$)
 δ [ppm] = 8.67 (br s, 1H, NH3), 8.03 (d, 2H, benzoyl-H$_A$, J = 7.01), 7.61 (t, 1H, benzoyl-H$_C$, J = 7.31), 7.43 (t, 2H, benzoyl-H$_B$, J = 7.31), 7.22 (s, 1H, H6), 6.29 (dd, 1H, H1', J = 6.09 and J = 7.92), 4.81 (s, 2H, -O-C$\underline{H}_2$-O-), 4.62 (m, 2H, H5'), 4.51 (m, 1H, H3'), 4.36 (q, 1H, H4', J = 3.35), 3.81 (t, 2H, -O-C$\underline{H}_2$, J = 6.09), 2.64 (t, 2H, C$\underline{H}_2$CN, J = 6.09), 2.59 – 2.10 (m, 2H, H2'), 1.67 (s, 3H, CH$_3$).

^{13}C-NMR: (100 MHz, $CDCl_3$)
 δ [ppm] = 166.2 (benzoyl-C=O), 163.5 (C4), 150.2 (C2), 134.8 (C6), 133.8

(benzoyl-C$_C$), 129.7 (benzoyl-C$_A$), 129.5 (benzoyl-C$_{quart}$), 128.9 (benzoyl-C$_B$), 117.7 (CN), 111.5 (C5), 94.7 (-O-$\underline{C}$H$_2$-O-), 85.2 (C1'), 82.8 (C4'), 77.4 (C3'), 64.2 (C5'), 63.2 (-O-$\underline{C}$H$_2$), 38.4 (C2'), 19.1 ($\underline{C}$H$_2$CN), 12.4 (CH$_3$).

ESI(+)-MS: m/z 430.1 [M + H$^+$]

3'-O-[(2-Cyanoethoxy)methyl]-2'-deoxythymidine

86
C$_{14}$H$_{19}$N$_3$O$_6$
325.32 g/mol

1.2 g (2.8 mmol) of compound **85** were dissolved in 30 ml MeOH and treated with 20 ml 32 % aq. ammonia. The solution was stirred at room temperature overnight, concentrated and the resulting residue purified by column chromatography with CH$_2$Cl$_2$/MeOH (5 – 10 % MeOH) to give the product as yellow glassy oil.

Yield: 708 mg (2.18 mmol, 78 %)
TLC: R$_f$ = 0.33 (CH$_2$Cl$_2$/MeOH: 9/1)

^{1}H-NMR: (400 MHz, CDCl$_3$)
δ [ppm] = 8.97 (br s, 1H, NH3), 7.47 (s, 1H, H6), 6.18 (ψt, 1H, H1'), 4.77 (s, 2H, -O-C$\underline{H}$$_2$-O-), 4.52 (m, 1H, H3'), 4.08 (q, 1H, H4', J = 3.03), 3.89 (m, 2H, H5'), 3.78 (t, 2H, -O-C$\underline{H}$$_2$, J = 6.06), 2.66 (t, 2H, C$\underline{H}$$_2$CN, J = 6.32), 2.36 (m, 2H, H2'), 1.90 (s, 3H, CH$_3$).

^{13}C-NMR: (100 MHz, CDCl$_3$)
δ [ppm] = 163.9 (C4), 150.6 (C2), 136.7 (C6), 118.2 (CN), 111.3 (C5), 94.4 (-O-$\underline{C}$H$_2$-O-), 86.3 (C1'), 85.4 (C4'), 77 (C3'), 63 (C5'), 62.4 (-O-$\underline{C}$H$_2$), 37.9 (C2'), 19.2 ($\underline{C}$H$_2$CN), 12.6 (CH$_3$).

ESI(+)-MS: m/z 325.9 [M + H$^+$]

3'-O-[(2-Cyanoethoxy)methyl]-2'-deoxythymidine-5'-phosphate

88
C$_{14}$H$_{20}$N$_3$O$_9$P
405.30 g/mol

150 mg (0.46 mmol) of nucleoside **86** were coevaporated three times with dry pyridine and dried overnight in vacuum. Under argon, the nucleoside was dissolved in 3 ml dry dioxane and 1.5 ml abs. pyridine, then 121 mg (0.6 mmol) of 2-chloro-1,3,2-benzodioxaphosphorin-4-one, dissolved in 1 ml dry dioxane were added within 2 min. The clear solution was stirred for 30 min at room temperature, the reaction was subsequently stopped by the addition of 0.25 ml water and 0.16 ml triethylamine (1.15 mmol). After solvent removal, the oily residue was lyophilized over four days. The resulting H-phosphonate **87** was dissolved in 50 ml dry pyridine, treated with 5.82 ml (46 mmol) chlorotri-methylsilane and stirred at room temperature for 5 min. A freshly prepared solution of 420 mg (1.38 mmol) iodine in 10 ml abs. pyridine was added in one lot and stirring continued for 30 min. The solvents were evaporated, the crude residue was diluted again in 5 ml pyridine and treated with 1 ml triethylamine and 2 ml water with stirring for a few minutes. After concentrating the mixture under reduced pressure, the residue was taken up in 8 ml water and purified in two portions on RP-FPLC and RP-HPLC under the conditions given below. The crude product obtained as yellow glassy oil was precipitated as sodium salt by employing the following procedure: The resulting 18 mg (0.044 mmol) of monophosphate **88** were rendered anhydrous by freeze-drying over three days. Under argon, the nucleotide was dissolved at 4 °C in 440 µl abs. MeOH to give a water-free methanolic 0.1 M monophosphate solution. A freshly prepared anhydrous sodium perchlorate solution, made from 82 mg (0.67 mmol) sodium perchlorate in 2 ml abs. acetone was added dropwise at 4 °C, and subsequently product **88** precipitated as white sodium salt. The solid was centrifuged (5 min, 4 °C, 12000

rpm), the liquid layer was transfused and the white residue dried in vacuum.

Yield: 17 mg (0.042 mmol as Na$^+$-salt, 9 %)

RP-FPLC: Method FPLC-3, Retention time 32 - 34 min
RP-HPLC: Method D, Retention time 18.36 min

^{1}H-NMR: (400 MHz, D$_2$O)
δ [ppm] = 7.78 (s, 1H, H6), 6.35 (dd, 1H, H1', J = 6.26 and J = 7.83), 4.92 (s, 2H, -O-C$\underline{H}_2$-O), 4.60 (m, 1H, H3'), 4.35 (m, 1H, H4'), 4.10 (m, 2H, H5'), 3.91 (t, 2H, -O-C$\underline{H}_2$, J = 6.26), 2.85 (t, 2H, C$\underline{H}_2$CN, J = 6.26), 2.53 - 2.37 (m, 2H, H2'), 1.93 (s, 3H, CH$_3$).

^{13}C-NMR: (100 MHz, D$_2$O)
δ [ppm] = 166.4 (C4), 151.6 (C2), 137.1 (C6), 119.5 (CN), 111.6 (C5), 94.3 (-O-$\underline{C}$H$_2$-O), 84.8 (C1'), 83.7 (C4'), 77.7 (C3'), 64.8 (-O-$\underline{C}$H$_2$), 63 (C5'), 36.7 (C2'), 18.1 ($\underline{C}$H$_2$CN), 11.4 (CH$_3$).

^{31}P-NMR: (162 MHz, D$_2$O)
δ [ppm] = 0.05 (s, monophosphate).

ESI(-)-MS: H-phosphonate **87**   m/z   388.1 [M - H$^+$]
            phosphate **88**   m/z   404.2 [M - H$^+$]

3N-Benzoyl-2'-deoxythymidine

90
C$_{17}$H$_{18}$N$_2$O$_6$
346.33 g/mol

6.06 g (25 mmol) of compound **81**, which were rendered anhydrous by repeated coevaporation with dry pyridine before, were dissolved in a mixture of 50 ml abs. pyridine, 21.8 ml (125 mmol) N-diisopropylethylamine and 8 ml (62.5 mmol) chlorotrimethylsilane. Stirring the solution under argon for about 30 min at room temperature led to complete silylation of the nucleoside. Subsequently 4.4 ml (37.5

mmol) of benzoyl chloride were added and stirring continued for 1 h until no starting material was detected by TLC. The reaction was stopped by the addition of 20 g KH_2PO_4 and 100 ml ice water on cooling then. After stirring the mixture a few minutes, colorless crystals precipitated which were collected by filtration, washed with 200 ml cold water and dried in vacuum. The crude solid was dissolved in 200 ml CH_2Cl_2/MeOH 1/1 and treated with 3.5 ml trifluoroacetic. The solution was stirred for 30 min at room temperature until all starting material was desilylated. The reaction mixture was filled into a separatory funnel then and washed once with 100 ml aq. 5 % $NaHCO_3$-solution. The aqueous layer was extracted three times with 100 ml CH_2Cl_2 each and the combined organic layers were dried over $MgSO_4$. After filtration and evaporation of the solvent, the crude product was purified by column chromatography using CH_2Cl_2/MeOH (5 - 10 % MeOH) as eluent.

Yield: 7.2 g (20.8 mmol, 83 %)
TLC: R_f = 0.29 (CH_2Cl_2/MeOH: 9/1)

^{1}H-NMR: (400 MHz, DMSO-d_6)
δ [ppm] = 7.97 (dd, 2H, benzoyl-H_A, J = 7.32 and J = 8.42), 7.78 (t, 1H, benzoyl-H_C, J = 7.46), 7.60 (t, 2H, benzoyl-H_B, J = 7.63), 6.15 (ψt, 1H, H1', J = 6.57), 5.26 (d, 1H, 3'-OH, J = 4.29), 5.11 (t, 1H, 5'-OH, J = 5.07), 4.29 (m, 1H, H3'), 3.81 (m, 1H, H4'), 3.62 (m, 2H, H5'), 2.24 - 2.14 (m, 2H, H2'), 1.86 (s, 3H, CH_3).

^{13}C-NMR: (100 MHz, DMSO-d_6)
δ [ppm] = 169.6 (benzoyl-C=O), 162.5 (C4), 149 (C2), 137.1 (C6), 135.5 (benzoyl-C_C), 131.1 (benzoyl-C_{quart}), 130.3 (benzoyl-C_A), 129.5 (benzoyl-C_B), 109.3 (C5), 87.6 (C4'), 84.5 (C1'), 70.2 (C3'), 61.1 (C5'), 39.6 (C2'), 12.2 (CH_3).

ESI(+)-MS: m/z 346.9 [M + H$^+$]

3N-Benzoyl-2'-deoxy-5'-O-(4-monomethoxytrityl)thymidine

91
$C_{37}H_{34}N_2O_7$
618.68 g/mol

3.46 g (10 mmol) of well-dried nucleoside **90** were dissolved in 40 ml abs. pyridine under argon and treated with 4.14 g (13 mmol) of *p*-monomethoxy-trityl chloride. The yellow solution was stirred at room temperature overnight, then the reaction was stopped by addition of 10 ml MeOH then and stirring for another 30 min. After solvent removal, the resulting oil was diluted in 200 ml CH_2Cl_2, washed three times with 100 ml water each, dried over $MgSO_4$, filtered and concentrated. The residue was purified on a silica gel column using CH_2Cl_2/MeOH (2 - 5 % MeOH) to furnish the product as yellow foam.

Yield: 4.85 g (7.8 mmol, 78 %)
TLC: R_f = 0.36 (CH_2Cl_2/MeOH: 95/5)

^{1}H-NMR: (400 MHz, $CDCl_3$)
δ [ppm] = 7.93 (m, 2H, benzoyl-H_A), 7.70 (s, 1H, H6), 7.63 (t, 1H, benzoyl-H_C, *J* = 7.36), 7.48 (t, 2H, benzoyl-H_B, *J* = 7.58), 7.44 - 7.27 (m, 12H, MMT), 6.87 (d, 2H, MMT, *J* = 8.84), 6.38 (dd, 1H, H1', *J* = 6.06 and *J* = 7.33), 4.59 (m, 1H, H3'), 4.04 (q, 1H, H4', *J* = 3.03), 3.81 (s, 3H, -OCH_3), 3.45 (ddd, 2H, H5', *J* = 3.03, *J* = 10.61 and *J* = 32.59), 2.39 (m, 2H, H2'), 1.49 (s, 3H, CH_3).

^{13}C-NMR: (100 MHz, $CDCl_3$)
δ [ppm] = 169.2 (benzoyl-C=O), 163.0 (C4), 159.0 (MMT-C-OCH_3), 149.4 (C2), 143.9 (MMT), 135.6 (C6), 135.1 (benzoyl-C_C), 134.9 (MMT), 131.1 (benzoyl-C_{quart}), 130.7 (benzoyl-C_A), 130.5 (MMT), 129.3 (benzoyl-C_B), 128.5 (MMT), 128.2 (MMT), 127.5 (MMT), 113.5 (MMT), 111.5 (C5), 86.3 (C4'), 85.1 (C1'), 72.5 (C3'), 63.7 (C5'), 55.4 (MMT-OCH_3), 41.3 (C2'), 12 (CH_3).

ESI(-)-MS: *m/z* 641.2 [M + Na$^+$]

Elemental analysis: calculated: C, 71.83; H, 5.54; N, 4.53; O, 18.10
found: C, 71.81; H, 5.59; N, 4.55

3N-Benzoyl-3'-O-(2-cyanoethyl)-2'-deoxy-5'-O-(4-monomethoxytrityl)thymidine

92
$C_{40}H_{37}N_3O_7$
671.74 g/mol

In a well-dried sealable Erlenmeyer flask with magnetic stirrer, 1.24 g (2 mmol) of compound **91** was dissolved in 2.7 ml (40 mmol) freshly distilled acrylonitrile and 5 ml *tert*-butanol with stirring. After complete dilution, 652 mg (2 mmol) of cesium carbonate were added and the pale yellow suspension was vigorously stirred for about 2.5 h at room temperature under argon. The reaction mixture was diluted with 100 ml CH_2Cl_2 and filtered through Celite. The filtrate was concentrated and the residue purified on a silica gel column with CH_2Cl_2/MeOH (0 - 1 % MeOH), yielding the product as pale yellow foam.

Yield: 1.23 g (1.83 mmol, 92 %)
TLC: R_f = 0.43 (CH_2Cl_2)

^{1}H-NMR: (400 MHz, $CDCl_3$)
δ [ppm] = 7.93 (dd, 2H, benzoyl-H_A, J = 7.07 and J = 8.59), 7.71 (s, 1H, H6), 7.64 (t, 1H, benzoyl-H_C, J = 7.58), 7.49 (t, 2H, benzoyl-H_B, J = 7.58), 7.43 - 7.26 (m, 12H, MMT), 6.87 (d, 2H, MMT, J = 8.84), 6.31 (dd, 1H, H1', J = 5.81 and J = 8.08), 4.24 (m, 1H, H3'), 4.15 (q, 1H, H4', J = 2.78), 3.81 (s, 3H, OCH_3), 3.63 (t, 2H, -O-C$\underline{H}_2$, J = 6.32), 3.54 - 3.33 (m, 2H, H5'), 2.55 (t, 2H, C$\underline{H}_2$CN, J = 6.32), 2.51 (m, 1H, H2'), 2.28 (m, 1H, H2'), 1.51 (s, 3H, CH_3).

^{13}C-NMR: (100 MHz, $CDCl_3$)
δ [ppm] = 169.2 (benzoyl-C=O), 162.9 (C4), 159.0 (MMT-$\underline{C}$-OMe), 149.4 (C2), 143.8 (MMT-C_{quart}), 135.4 (C6), 135.1 (benzoyl-C_C), 134.8 (C2, C2', trityl), 131.1 (benzoyl-C_{quart}), 130.6 (benzoyl-C_A), 130.5 (MMT), 129.8 (MMT), 129.3 (benzoyl-C_B), 128.5 (MMT), 128.3 (MMT), 127.6 (MMT), 117.5 (CN), 113.5 (MMT), 111.4 (C5), 87.5 (C1'), 85.1 (C4'), 80.5 (C3'), 64.1 (-O-$\underline{C}H_2$), 63.7 (C5'), 55.4 (MMT-O$\underline{C}H_3$), 37.9 (C2'), 19.1 ($\underline{C}H_2$CN), 12.0 (CH_3).

MALDI(+)-MS: *m/z* 694.79 [M + Na$^+$]

Elemental analysis: calculated: C, 71.52; H, 5.55; N, 6.26; O, 16.67
                     found: C, 71.24; H, 5.54; N, 6.35

3N-Benzoyl-3'-O-(2-cyanoethyl)-2'-deoxythymidine

93
$C_{20}H_{21}N_3O_6$
399.40 g/mol

671 mg (1 mmol) of compound **92** were dissolved in 20 ml of a solution of 10 % p-toluenesulfonic acid in CH_2Cl_2/EtOH 1/1. The reaction mixture was stirred at room temperature for 2 h until all starting material was consumed. The solution was treated with 100 ml aq. sat. $NaHCO_3$-solution then and the organic layer separated from the aqueous layer. The aqueous layer was extracted three times with 100 ml CH_2Cl_2 each and the combined organic layers were dried over $MgSO_4$, filtered and concentrated. The resulting colorless oil was purified on column chromatography with CH_2Cl_2/MeOH 95/5 to give the product as colorless foam.

Yield: 399 mg (1 mmol, 100 %)
TLC: R_f = 0.67 (CH_2Cl_2/MeOH: 9/1)

^{1}H-NMR: (250 MHz, DMSO-d_6)
δ [ppm] = 8.0 – 7.56 (m, 6H, H6, benzoyl-H), 6.11 (t, 1H, H1'), 5.20 (t, 1H, 3'-OH), 4.19 (m, 1H, H3'), 3.97 (m, 1H, H4'), 3.64 (t, 2H, -O-C$\underline{H}_2$, J = 6.04), 3.63 (m, 2H, H5'), 2.78 (t, 2H, C$\underline{H}_2$CN, J = 6.04), 2.35 – 2.28 (m, 2H, H2'), 1.87 (s, 3H, CH$_3$).

^{13}C-NMR: (63 MHz, DMSO-d_6)
δ [ppm] = 169.6 (benzoyl-C=O), 162.5 (C4), 149.0 (C2), 137.0 (C6), 135.5 (benzoyl-C_C), 131.1 (benzoyl-C_{quart}), 130.4 (benzoyl-C_A), 129.5 (benzoyl-C_B), 119.2 (CN), 109.5 (C5), 84.9 (C1'), 84.5 (C4'), 70.2 (C3'), 63.5 (-O-$\underline{C}H_2$), 61.4 (C5'), 36.5 (C2'), 18.3 ($\underline{C}H_2$CN), 12.2 (CH$_3$).

ESI(+)-MS: m/z   399.9 [M + H$^+$]

Elemental analysis: calculated: C, 60.14; H, 5.30; N, 10.52; O, 24.04
 found: C, 60.39; H, 5.47; N, 10.66

3'-O-(2-Cyanoethyl)-2'-deoxythymidine

94
$C_{13}H_{17}N_3O_5$
295.29 g/mol

300 mg (0.75 mmol) of nucleoside **93** were dissolved in 10 ml MeOH and cooled down to 0 °C. To this solution, 3 ml of 27 % aq. ammonia were added dropwise and the mixture was stirred for 30 min at 0 °C, then 1.5 h at room temperature. After complete conversion of the starting material the reaction mixture was concentrated. The residual oil was purified on a silica gel column with CH_2Cl_2/MeOH 95/5 to furnish the product as colorless crystals.

Yield: 212 mg (0.72 mmol, 96 %)
TLC: R_f = 0.30 (CH_2Cl_2/MeOH: 9/1)

^{1}H-NMR: (250 MHz, DMSO-d_6)
 δ [ppm] = 11.31 (br s, 1H, H3), 7.68 (s, 1H, H6), 6.12 (ψt, 1H, H1'), 5.12 (m, 1H, 5'-OH, J = 5.38), 4.14 (m, 1H, H3'), 3.92 (m, 1H, H4'), 3.63 (t, 2H, O-C$\underline{H}_2$, J = 6.04), 3.58 (m, 2H, H5'), 2.78 (t, 2H, C$\underline{H}_2$CN, J = 6.04), 2.28 – 2.09 (m, 2H, H2'), 1.77 (s, 3H, CH$_3$).

^{13}C-NMR: (63 MHz, DMSO-d_6)
 δ [ppm] = 163.7 (C4), 150.5 (C2), 136.0 (C6), 119.2 (CN), 109.6 (C5), 84.5 (C1'), 83.7 (C4'), 79.4 (C3'), 63.5 (-O-$\underline{C}H_2$), 61.4 (C5'), 36.2 (C2'), 18.3 ($\underline{C}H_2$CN), 12.3 (CH$_3$).

ESI(+)-MS: *m/z* 318.0 [M + Na$^+$]

Elemental analysis: calculated: C, 52.88; H, 5.80; N, 14.23; O, 27.09
 found: C, 53.07; H, 5.91; N, 14.23

3'-O-(2-Cyanoethyl)-2'-deoxythymidine-5'-phosphate

95
$C_{13}H_{18}N_3O_8P$
375.27 g/mol

385 mg (1.3 mmol) of nucleoside **94** were dissolved in 10 ml dry trimethyl-phosphate and cooled down to 0 °C. The solution was treated with 418 mg (1.95 mmol) 1,8-bis-(dimethylamino)naphthalene and, after stirring for 5 min, 0.16 ml (1.7 mmol) of phosphorus oxychloride were added dropwise on cooling. The pale yellow clear solution was stirred for 3 h at 0 °C and immediately quenched by addition of 20 ml cold triethylammonium bicarbonate buffer (pH 7) with agitating the mixture for 1 h more. The aqueous layer was extracted five times with 100 ml methyl *tert*-butyl ether for removal of the proton sponge and of trimethyl phosphate residues. The aqueous layer was lyophilized and purified on RP-HPLC. The crude product was obtained as colorless glassy oil which was characterized via NMR- and mass spectrometry. For easier handling and storage of the product, it was precipitated as sodium salt: The resulting 250 mg (0.66 mmol) of monophosphate **95** were rendered anhydrous by freeze-drying over three days. Under argon, the nucleotide was dissolved at 4 °C in 6.6 ml abs. MeOH to give a water-free methanolic 0.1 M monophosphate solution. A freshly prepared anhydrous sodium perchlorate solution (made from 1.21 g (9.9 mmol) sodium perchlorate in 33 ml abs. acetone) was added dropwise at 4°C causing the sodium salt's precipitation. The precipitation was centrifuged (5 min, 4 °C, 12000 rpm), the liquid layer transfused and the colorless crystals of monophosphate **95** were isolated and dried in vacuum.

Yield: 251 mg (0.67 mmol, 52 %)

RP-HPLC: Method B, Retention time: 10.88 min

^{1}H-NMR: (400 MHz, D$_2$O)

δ [ppm] = 7.69 (s, 1H, H6), 6.22 (ψt, 1H, H1'), 4.31 (m, 1H, H3'), 4.22 (m, 1H, H4'), 3.95 (m, 2H, H5'), 3.72 (t, 2H, O-C<u>H</u>₂, J = 6.04), 2.71 (t, 2H, C<u>H</u>₂CN, J = 6.04), 2.39 (m, 1H, H2'), 2.23 (m, 1H, H2'), 1.81 (s, 3H, CH₃).

^{13}C-NMR: (100 MHz, D₂O)
δ [ppm] = 166.3 (C4), 151.5 (C2), 137.2 (C6), 119.6 (CN), 111.6 (C5), 85.0 (C1'), 83.5 (C4'), 80.0 (C3'), 64.9 (O-<u>C</u>H₂), 63.7 (C5'), 36.2 (C2'), 18.3 (<u>C</u>H₂CN), 11.5 (CH₃).

^{31}P-NMR: (162 MHz, D₂O)
δ [ppm] = 2.11 (s, monophosphate).

MALDI(-)-MS: m/z   373.8 [M - 2H⁺]

3'-O-(2-Cyanoethyl)-2'-deoxythymidine-5'-(N,N-diisopropyl)phosphoramidite

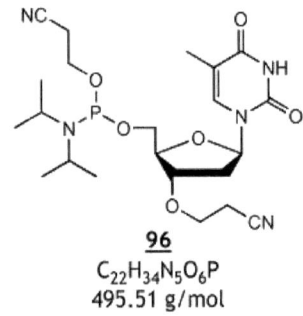

96
C₂₂H₃₄N₅O₆P
495.51 g/mol

236 mg (0.8 mmol) of the well-dried nucleoside **94** were dissolved under argon in 8 ml abs. CH₂Cl₂ and treated with 381 μl (1.2 mmol) (2-cyanoethyl)-di(N,N-diisopropyl)phosphine and 110 mg (0.92 mmol) 4,5-dicyanoimidazole. The yellow clear solution was stirred at room temperature for 4 h, then diluted with 20 ml CH₂Cl₂ and washed once with 25 ml aq. sat. NaHCO₃. The organic layer was separated, dried over Na₂SO₄, filtered and concentrated. The crude product was purified on a short silica gel column under slight positive argon pressure. The pure product was obtained as glassy oil (containing both isomers of **94**) which was stored at -20 °C under argon.

Yield: 328 mg (0.66 mmol, 83 %)
TLC: R$_f$ = 0.4 (CH₂Cl₂/MeOH: 9/1)

^{1}H-NMR: (400 MHz, acetone-d₆)

δ [ppm] = 9.88 (br s, 2H, NH), 7.69 (s, 1 H, H6), 7.56 (s, 1 H, H6), 6.31 (dd, 1 H, H1', J = 5.83, J = 8.38), 6.25 (dd, 1 H, H1', J = 5.60, J = 8.60), 4.36 (m, 1 H, H3'), 4.32 (m, 1 H, H3'), 4.21 (m, 2 H, H4'), 4.0 – 3.63 (m, 10 H, H5', phosphine-iso-propyl-H, phosphine-O-C$\underline{H}_2$), 3.91 (t, 2 H, H8, J = 6.10), 3.81 (t, 2 H, H8, J = 6.10), 2.80 (t, 2 H, phosphine-C$\underline{H}_2$CN, J = 5.81), 2.77 (t, 2 H, H9, J = 6.06), 2.41 – 2.26 (m, 4 H, H2'), 1.87 (s, 3 H, CH$_3$), 1.86 (s, 3 H, CH$_3$), 1.28 – 0.99 (m, 24 H, phosphine-iso-propyl-H).

^{13}C NMR: (100 MHz, acetone-d_6)
δ [ppm] = 163.2 (C4), 150.2 (C2), 135.4 (C6), 135.2 (C6), 118.2 (phosphine-CN), 118 (CN), 110 (C5), 84.6 (C1'), 84.3 (C1'), 83.5 (C4'), 80.1 (C3'), 63.8 (-O-$\underline{C}$H$_2$), 63.2 (-O-$\underline{C}$H$_2$), 61.4 (C5'), 58.4 (phosphine-O-$\underline{C}$H$_2$), 42.7 (phosphine-$\underline{C}$-iso-propyl), 36.6 (C2'), 23.9 (phosphine-$\underline{C}$-iso-propyl), 19.8 (phosphine-CH$_2$), 18.0 ($\underline{C}$H$_2$CN), 11.5 (CH$_3$)

^{31}P NMR: (75 MHz, acetone-d_6)
δ [ppm] = 148.8 (s, P-α), 148.2 (s, P-α).

ESI(-)-MS: m/z   494.4 [M - H$^+$]

2'-Deoxythymidine-5'-phosphate

97
C$_{10}$H$_{15}$N$_2$O$_8$P
322.21 g/mol

315 mg (1.3 mmol) of 2'-deoxythymidine were dissolved under argon in 10 ml abs. trimethylphosphate and cooled down to -4 °C. At that temperature, 0.25 ml (2.6 mmol) of phosphorus oxychloride were added dropwise and the reaction mixture was stirred for 6 h at -4 °C. The reaction was stopped by the addition of 20 ml TEAB buffer (pH 7) and stirring continued for 30 min. The aqueous layer was extracted five times with 100 ml methyl tert-butyl ether each, lyophilized and purified on RP-HPLC. The crude product was obtained as colorless glassy oil which was characterized via NMR- and mass spectrometry. For easier handling and storage of the product, it was precipitated as

sodium salt: The resulting 213 mg (0.66 mmol) of monophosphate **97** were rendered anhydrous by freeze-drying over three days. Under argon, the nucleotide was dissolved at 4 °C in 6.6 ml abs. MeOH to give a water-free methanolic 0.1 M monophosphate solution. A freshly prepared anhydrous sodium perchlorate solution (made from 1.21 g (9.9 mmol) sodium perchlorate in 33 ml abs. acetone) was added dropwise at 4 °C causing precipitation of the nucleotide as colorless crystals. The precipitated sodium salt of **97** was centrifuged (5 min, 4 °C, 12000 rpm), the liquid layer was transfused and the product dried.

Yield: 287 mg (0.89 mmol, 69 %)

RP-HPLC: Method C, Retention time 8.57 min

^{1}H-NMR: (400 MHz, D$_2$O)
δ [ppm] = 7.58 (s, 1H, H6), 6.15 (ψt, 1H, H1'), 4.37 (m, 1H, H3'), 3.97 (m, 1H, H4'), 3.87 (m, 2H, H5'), 2.16 (m, 2H, H2'), 1.71 (s, 3H, CH$_3$).

^{13}C-NMR: (400 MHz, D$_2$O)
δ [ppm] = 166.4 (C2), 151.5 (C4), 137.3 (C6), 111.5 (C5), 85.4 (C1'), 84.9 (C4'), 71.0 (C3'), 64.6 (C5'), 38.6 (C2'), 11.5 (CH$_3$).

^{31}P-NMR: (162 MHz, D$_2$O)
δ [ppm] = 0.66 (s, monophosphate).

MALDI(+)-MS: m/z   362.13 [M + K$^+$ + 2H$^+$]

<u>4N-(N,N-Dimethylaminomethylidenyl)-2'-deoxyadenosine</u>

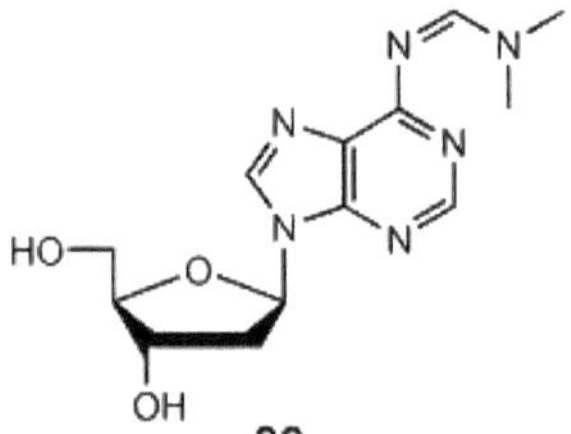

99
C$_{13}$H$_{18}$N$_6$O$_3$
306.32 g/mol

7.54 g (30 mmol) of 2'-deoxyadenosine were dissolved in 40 ml dry DMF and treated with

41 ml (300 mmol) of N,N-dimethylformamide dimethyl acetal. The solution was heated for 3 h at 55 °C and concentrated under reduced pressure then. The residue was purified on silica gel column with CH_2Cl_2/MeOH (10 - 20 % MeOH) as eluent to give the product as colorless crystals.

Yield: 7.23 g (23.6 mmol, 79 %)
TLC: R_f = 0.05 (CH_2Cl_2/MeOH: 9/1)

^{1}H-NMR: (400 MHz, DMSO-d_6)
δ [ppm] = 8.9 (s, 1H, formamidino-CH), 8.44 (s, 1H, H6), 8.40 (s, 1H, H2), 6.39 (dd, 1H, H1', J = 6.59, J = 7.32), 5.32 (d, 1H, 3'-OH, J = 4.12), 5.14 (t, 1H, 5'-OH, J = 5.16), 4.42 (m, 1H, H3'), 3.89 (m, 1H, H4'), 3.57 (m, 2H, H5'), 3.18 (s, 3H, CH_3), 3.12 (s, 3H, CH_3), 2.77 - 2.26 (m, 2H, H2').

^{13}C-NMR: (100 MHz, DMSO-d_6)
δ [ppm] = 159.4 (C4), 158.2 (formamidino-C=N), 151.9 (C2), 151.1 (C7a), 141.3 (C6), 125.8 (C4a), 88 (C4'), 83.9 (C1'), 71 (C3'), 61.9 (C5'), 40.8 (formamidino-CH_3), 39.5 (C2'), 34.6 (formamidino-CH_3).

ESI(-)-MS: m/z 305.1 [M - H$^+$]

4N-(N,N-Dimethylaminomethylidenyl)-5'-O-benzoyl-2'-deoxyadenosine

100
$C_{20}H_{22}N_6O_4$
410.43 g/mol

1.53 g (5 mmol) of compound **99** were coevaporated three times with dry pyridine and dried overnight in vacuum. The nucleoside was dissolved in 100 ml abs. pyridine and cooled down to - 20 °C. In a separate flask, 640 µl (5.5 mmol) of benzoyl chloride were diluted in 8 ml abs. pyridine, taken up with a syringe and added dropwise through a septum to the nucleoside within 1 h. After consumption of the starting material (roughly 2.5 h), the reaction was stopped by addition of 5 ml MeOH and stirring the mixture for

15 min. The solvent was removed under reduced pressure and the resulting syrup purified on silica gel column using CH_2Cl_2/MeOH (2 - 10 % MeOH) as eluent.

Yield: 1.66 g (4.04 mmol, 81 %)
TLC: R_f = 0.3 (CH_2Cl_2/MeOH: 9/1)

^{1}H-NMR: (400 MHz, DMSO-d_6)
δ [ppm] = 8.89 (s, 1H, formamidino-CH), 8.41 (s, 1H, H6), 8.39 (s, 1H, H2), 7.91 (ψd, 2H, benzoyl-H_A, J = 7.68), 7.65 (ψt, 1H, benzoyl-H_C, J = 7.32), 7.49 (ψt, 2H, benzoyl-H_B, J = 7.68), 6.43 (ψt, 1H, H1', J = 6.59), 5.60 (d, 1H, 3'-OH, J = 4.39), 4.64 (m, 1H, H3'), 4.57 - 4.40 (m, 2H, H5'), 4.15 (m, 1H, H4'), 3.19 (s, 3H, formamidino-CH_3), 3.12 (s, 3H, formamidino-CH_3), 2.97 - 2.35 (m, 2H, H2').

^{13}C-NMR: (100 MHz, DMSO-d_6)
δ [ppm] = 165.6 (benzoyl-C=O), 159.1 (C4), 158.1 (formamidino-C=N), 151.8 (C2), 151.1 (C7a), 141.5 (C6), 133.4 (benzoyl-C_C), 129.4 (benzoyl-C_{quart}), 129.2 (benzoyl-C_A), 128.7 (benzoyl-C_B), 125.8 (C4a), 84 (C4'), 83.5 (C1'), 70.5 (C3'), 64.5 (C5'), 40.7 (formamidino-CH_3), 38.4 (C2'), 34.6 (formamidino-CH_3).

ESI(+)-MS: m/z 411.3 [M + H$^+$]

<u>4N-(N,N-Dimethylaminomethylidenyl)-5'-O-benzoyl-3'-O-(2-cyanoethyl)-2'-deoxyadenosine</u>

101
$C_{23}H_{25}N_7O_4$
463.49 g/mol

820 mg (2 mmol) of nucleoside **100** were dissolved under argon in an Erlenmeyer flask with triangle stirrer bar in 6.6 ml (100 mmol) freshly distilled acrylonitrile and in 5 ml *tert*-butanol. After stirring the mixture for a few minutes, 652 mg (2 mmol) of cesium carbonate were added in one lot and vigorous stirring continued for 2 h. After completion of the reaction, the mixture was diluted in 150 ml CH_2Cl_2 and filtered

through Celite for removal of the insoluble carbonate. After concentration of the filtrate and purification of the residue on a silica gel column CH_2Cl_2/MeOH (5 - 10 % MeOH) as eluent, the product was obtained as yellowish foam.

Yield: 487 mg (1.05 mmol, 53 %)
TLC: R_f = 0.38 (CH_2Cl_2/MeOH: 9/1)

^{1}H-NMR: (400 MHz, $CDCl_3$)
δ [ppm] = 8.92 (s, 1H, formamidino-CH), 8.49 (s, 1H, H2), 8.0 (s, 1H, H6), 7.98 (ψd, 2H, benzoyl-H_A, J = 8.42), 7.56 (ψt, 1H, benzoyl-H_C, J = 7.32), 7.42 (ψt, 2H, benzoyl-H_B, J = 7.68), 6.4 (ψt, 1H, H1', J = 6.59), 4.61 (m, 2H, H5'), 4.51 (m, 1H), 4.44 (m, 1H), 3.77 (t, 2H, -O-C$\underline{H}_2$, J = 6.22), 3.25 (s, 3H, formamidino-CH_3), 3.19 (s, 3H, formamidino-CH_3), 3.06 (m, 1H, H2'), 2.66, (t, 2H, C$\underline{H}_2$CN, J = 6.22), 2.64 (m, 1H, H2').

^{13}C-NMR: (100 MHz, $CDCl_3$)
δ [ppm] = 166.4 (benzoyl-C=O), 159.9 (C4), 158.1 (formamidino-C=N), 152.8 (C2), 151.3 (C7a), 140.4 (C6), 133.5 (benzoyl-C_C), 129.8 (benzoyl-C_A), 129.6 (benzoyl-C_{quart}), 128.7 (benzoyl-C_B), 126.9 (C4a), 117.5 (CN), 85 (C4'), 82.5 (C1'), 80.5 (C3'), 64.6 (-O-$\underline{C}H_2$), 64.3 (C5'), 41.4 (formamidino-CH_3), 37.1 (C2'), 35.3 (formamidino-CH_3), 19.2 ($\underline{C}H_2$CN).

ESI(+)-MS: m/z   464.3 [M + H$^+$]

3'-O-(2-Cyanoethyl)-2'-deoxyadenosine

102
$C_{13}H_{16}N_6O_3$
304.30 g/mol

750 mg (1.62 mmol) of compound **101** were filled in a sealable bottle and dissolved in 20 ml MeOH. 50 ml of 32 % aq. ammonia were added and the mixture was agitated overnight at room temperature. After solvent removal, the residue was purified on a silica gel column with CH_2Cl_2/MeOH (5 - 10 % MeOH) to furnish the product as yellowish

crystals.

Yield: 318 mg (1.05 mmol, 65 %)
TLC: R_f = 0.18 (CH_2Cl_2/MeOH: 9/1)

^{1}H-NMR: (400 MHz, DMSO-d_6)
δ [ppm] = 8.34 (s, 1H, H6), 8.13 (s, 1H, H2), 7.31 (br s, 2H, NH_2), 6.3 (dd, 1H, H1', J = 5.48, J = 8.61), 5.37 (br s, 1H, 5'-OH), 4.31 (m, 1H, H3'), 4.03 (m, 1H, H4'), 3.68 (t, 2H, O-C$\underline{H}_2$, J = 6.26), 3.58 (m, 2H, H5'), 2.81 (t, 2H, C$\underline{H}_2$CN, J = 6.26), 2.82 - 2.46 (m, 2H, H2').

^{13}C-NMR: (100 MHz, DMSO-d_6)
δ [ppm] = 156.1 (C4), 152.4 (C2), 148.9 (C7a), 139.6 (C6), 129.2 (C4a), 119.3 (CN), 85.3 (C4'), 84 (C1'), 80 (C3'), 63.4 (-O-$\underline{C}H_2$), 61.9 (C5'), 36.2 (C2'), 18.3 ($\underline{C}H_2$CN).

MALDI(-)-MS: m/z   304.4

3'-O-(2-Cyanoethyl)-2'-deoxyadenosine-5'-phosphate

104
$C_{13}H_{17}N_6O_6P$
384.28 g/mol

84 mg (0.275 mmol) of well-dried nucleoside **102** were dissolved in 3 ml dry dioxane and 1.5 ml dry pyridine, then 73 mg (0.358 mmol) of 2-chloro-4H-1,3,2-benzodioxaphosphorinan-4-one in 1 ml anhydrous dioxane were added via syringe within 1 min. The mixture was stirred 30 min at room temperature, the reaction was immediately stopped by addition of 0.5 ml water and 0.2 ml TEA. After stirring the mixture for another 5 min, the solvents were removed under reduced pressure and the residue freeze dried over three days. The resulting H-phosphonate **103**, as identified by mass spectroscopy, was dissolved again in 50 ml abs. pyridine and treated with 3.48 ml (27.5 mmol) chlorotrimethylsilane under argon. After 5 min, a freshly prepared solution

of 251 mg (0.825 mmol) iodine in 10 ml anhydrous pyridine was added and stirring maintained for 30 min at room temperature. The solvents were removed under reduced pressure, the residue was taken up in 5 ml dry pyridine again and treated with 2 ml water for completion of the oxidation. The excess of iodine was reduced by adding 200 mg solid $Na_2S_2O_5$ to the solution, then the mixture was evaporated to dryness. The resulting crude material was diluted again in 4 ml Millipore water, filtered through syringe filter and submitted to RP-FPLC for the first purification step. After identification of the product containing fraction, it was concentrated and further purified on RP-HPLC to give the nucleotide as yellow oil. The resulting 17 mg (0.044 mmol) of monophosphate **104** were rendered anhydrous by freeze-drying over three days. Under argon, the nucleotide was dissolved at 4 °C in 0.44 ml abs. MeOH to give a water-free methanolic 0.1 M monophosphate solution. A freshly prepared anhydrous sodium perchlorate solution (made from 81 mg (0.664 mmol) sodium perchlorate in 2.2 ml abs. acetone) was added dropwise at 4 °C causing precipitation of the nucleotide as sodium salt. The crystalline product was centrifuged (5 min, 4 °C, 12000 rpm), the liquid layer was transfused and the product isolated and dried in vacuum.

Yield: 30 mg (0.078 mmol, 28 %)

RP-FPLC: Method FPLC-3, Retention time 28 - 30 min
RP-HPLC: Method D, Retention time 19.45 min

^{1}H-NMR: (400 MHz, D_2O)
δ [ppm] = 8.41 (s, 1H, H6), 8.14 (s, 1H, H2), 6.41 (dd, 1H, H1', J = 6.26 and J = 7.83), 4.54 (m, 1H, H3'), 4.42 (m, 1H, H4'), 4.05 (m, 2H, H5'), 3.89 (t, 2H, O-C$\underline{H}_2$, J = 6.26), 2.86 (t, 2H, C$\underline{H}_2$CN, J = 6.26), 2.75 (m, 2H, H2'), 1.81 (s, 3H, CH_3).

^{13}C-NMR: (100 MHz, D_2O)
δ [ppm] = 155.1 (C4), 152.3 (C2), 148.5 (C7a), 139.9 (C6), 119.9 (CN), 118.4 (C4a), 84 (C4'), 83.9 (C1'), 80.3 (C3'), 64.9 (C5'), 64 (O-$\underline{C}H_2$), 36.9 (C2'), 18.4 ($\underline{C}H_2$CN).

^{31}P-NMR: (162 MHz, D_2O)
δ [ppm] = 0.45 (s, monophosphate).

ESI(-)-MS: H-phosphonate **103**   m/z   367.1 [M - H$^+$]
             Phosphate   **104**   m/z   383.1 [M - H$^+$]

7.3 Oligonucleotide Synthesis

The 3'-phosphoramidites for the synthesis (direction 3' to 5') of the unmodified DNA 8mer were purchased from Pharmacia Biotech, the 5'-phosphoramidites for the inverse synthesis (direction 5' to 3') of the modified DNA 8mer from Glen Research. The oligodeoxynucleotides were synthesized on an Expedite Nucleic Acid Synthesis System using 500 Ångstrøm CPG columns from Applied Biosystems. The sequence of the unmodified reference oligomer is 3'-TATGCGGT-5', the sequence of the modified oligomer is 3'-ce-TATGCGGT-5' with ce = cyanoethyl. The deprotection test on the oligomer was performed in a heating block Thermomixer comfort from Eppendorf. OD measurements were carried out with a Hitachi U-1100 Spectrophotometer at λ = 260 nm. The oligomers were purified on an anion-exchange HPLC (Method E) and desalted with PD-10 columns (prepacked with Sephadex™ G-25 M) from GE Healthcare, dried under reduced pressure and characterized by MALDI-TOF mass using 6-Aza-2-thiothymine (ATT) / ammonium citrate matrix.

8 Annex

8.1 NMR and mass spectra

ESI(+) mass spectrum of nucleoside **4**

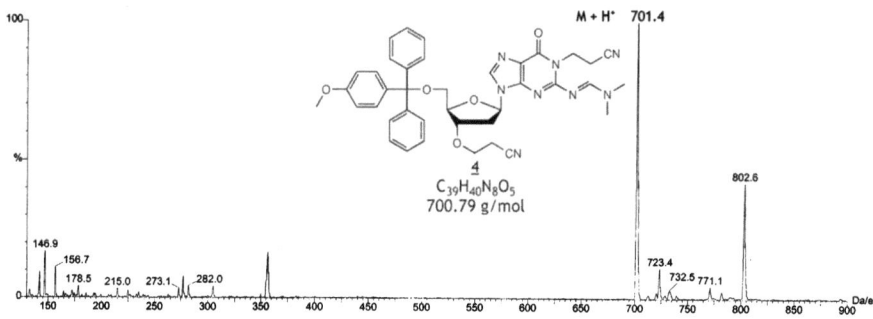

^{1}H-NMR spectrum of nucleoside **31** (DMSO-d_6, 400 MHz)

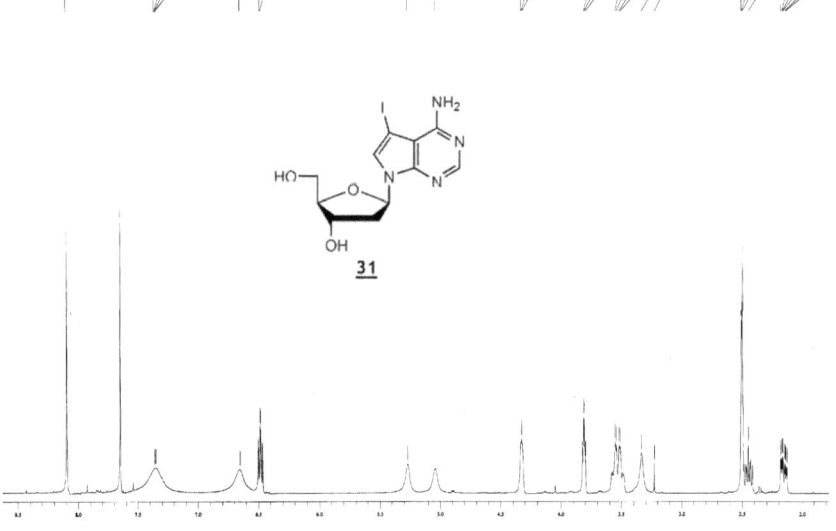

¹H-NMR spectrum of nucleoside **54** (DMSO-d_6, 400 MHz)

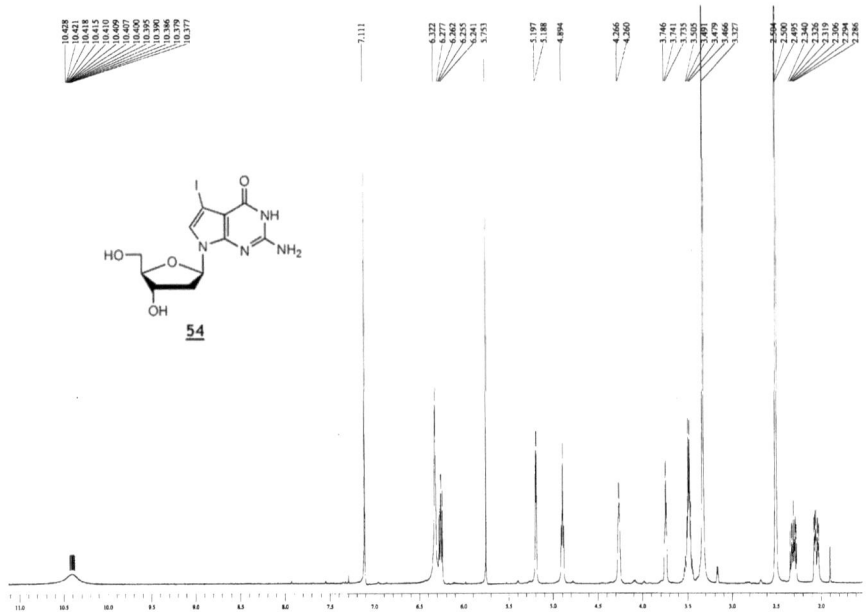

¹H-NMR spectrum of nucleoside **72** (DMSO-d_6, 250 MHz)

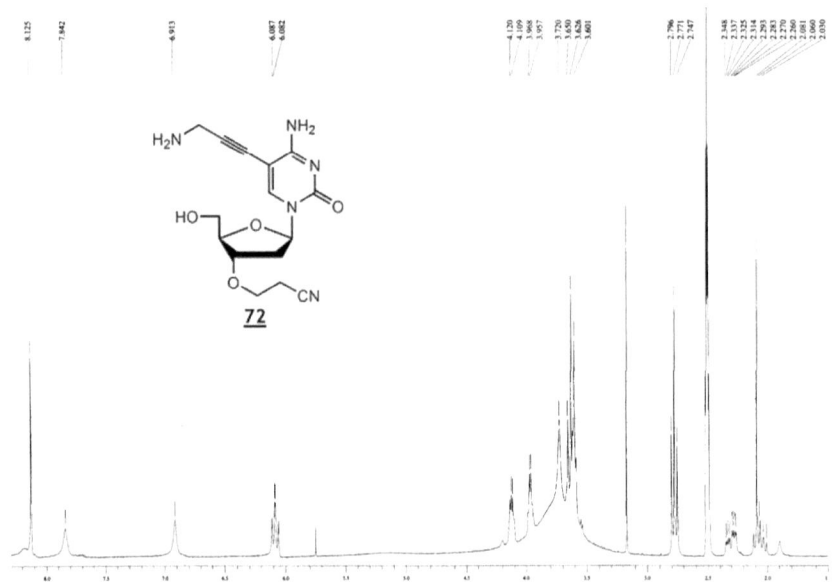

¹H-NMR spectrum of nucleoside **80** (DMSO-d_6, 300 MHz)

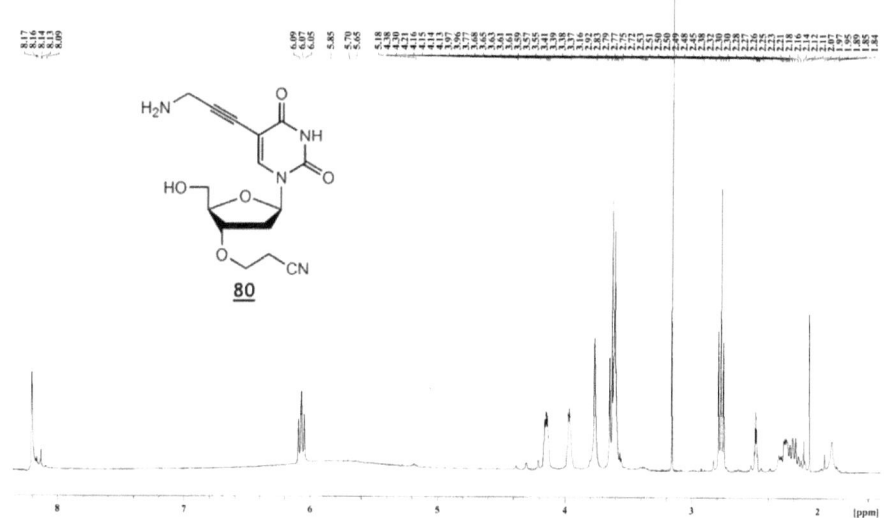

¹H-NMR spectrum of monophosphate **88** (D$_2$O, 300 MHz)

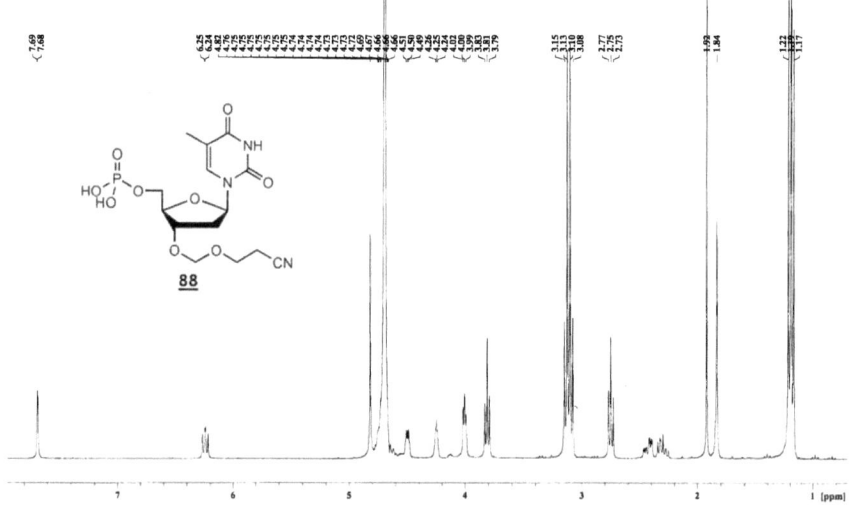

¹H-NMR spectrum of monophosphate 95 (D₂O, 400 MHz)

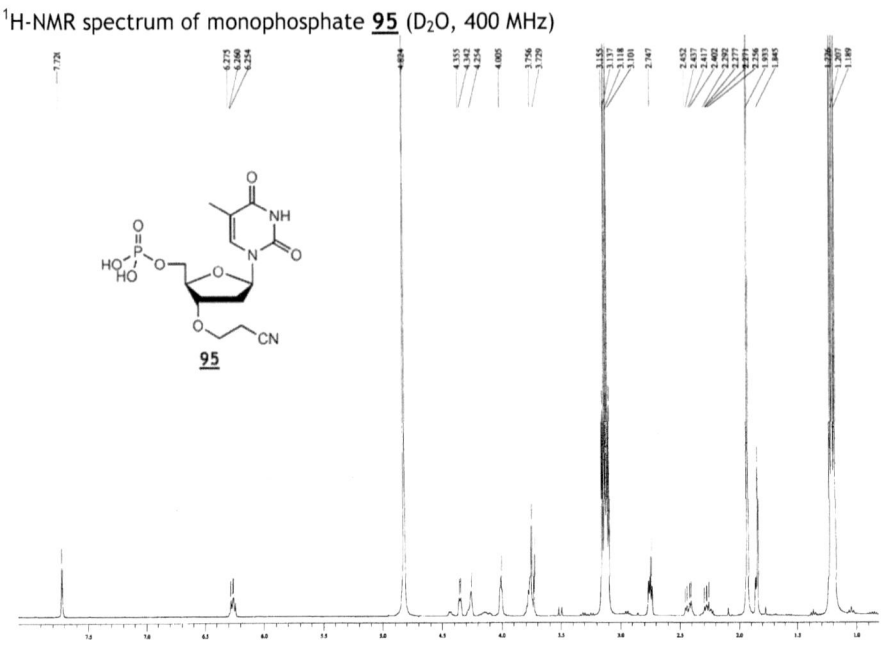

¹H-NMR spectrum of phosphoramidite 96 (acetone-d₆, 400 MHz)

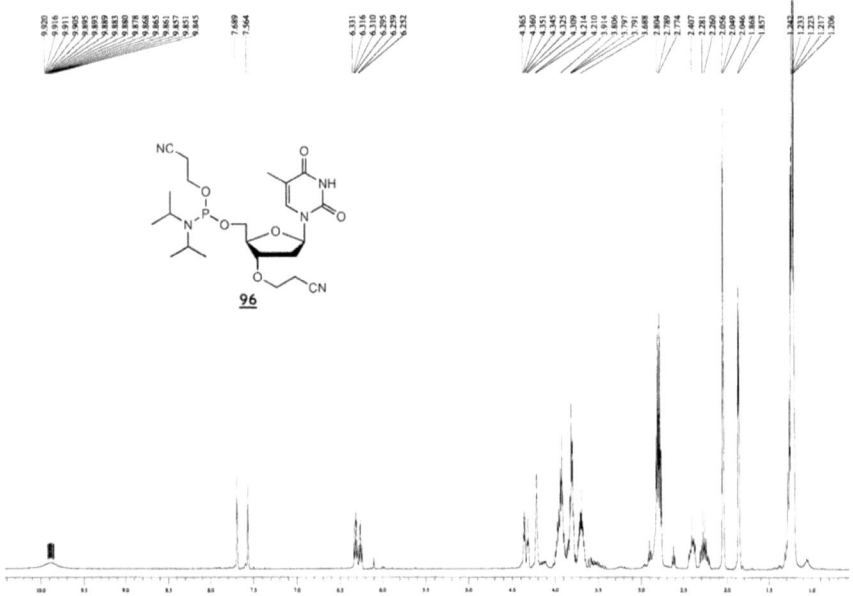

Annex
217

¹H-NMR spectrum of monophosphate **97** (D$_2$O, 400 MHz)

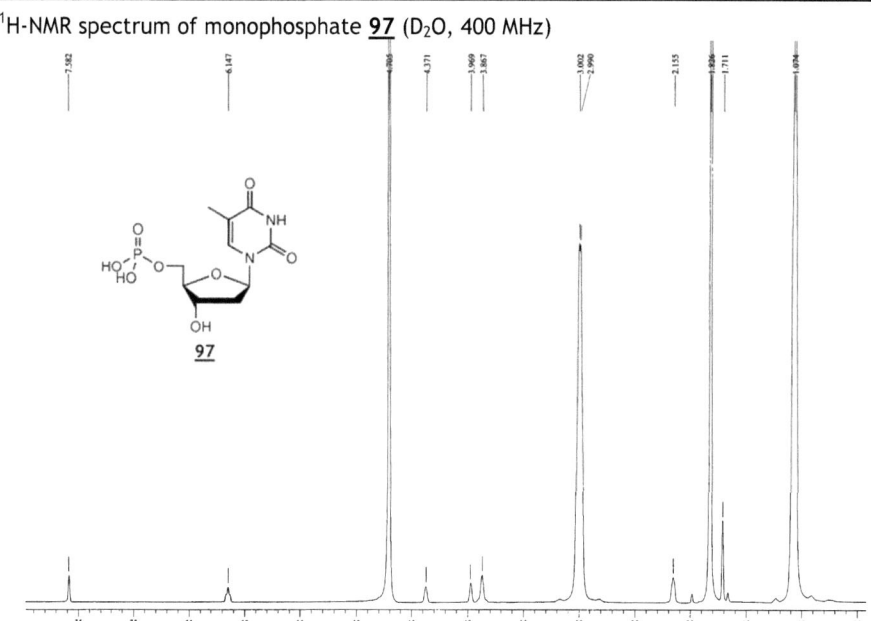

¹H-NMR spectrum of monophosphate **104** (D$_2$O, 400 MHz)

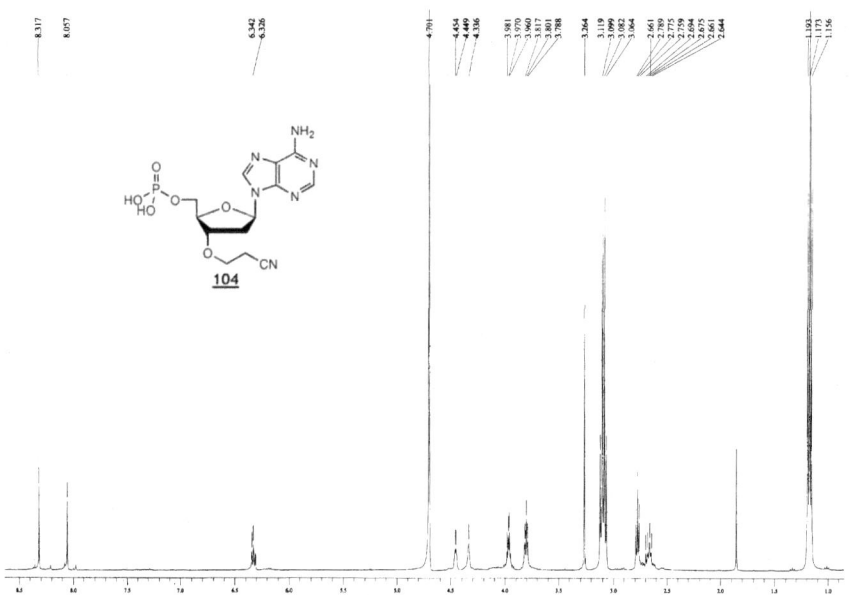

8.2 Abbreviations

A

	A	Adenosine
	abs.	absolute
	Ac	acetyl
	ACN	acetonitrile
	AcOH	acetic acid
	ant	anthranyloyl
	aq.	aqueous
	approx.	approximately
	Asp	asparagine

B

	BASS	Base Addition Sequencing Scheme
	bp	boiling point
	BSA	N,O-Bis(trimethylsilyl)acetamide
	Bz	benzoyl

C

	C	Cytosine
	CCD	charge-coupled device
	$CDCl_3$	chloroform-d_1
	CE	cyanoethyl
	CEM	cyanoethoxymethyl
	conc.	concentrated
	CRT	cyclic reversible termination

D

	D	duplet
	dAMP	2'-deoxyadenosine-5'-monophosphate
	DCE	1,2-dichloroethane
	DCM	dichloromethane (methylene chloride)
	ddNTP	dideoxynucleoside-5'-triphosphate (with N = A, C, G, U and T)
	DIPEA	N,N'-diisopropylethylamine
	dist.	distilled
	DMAP	4-(N,N'-dimethylamino)pyridine
	DMF	N,N-dimethylformamide
	DMSO	dimethylsulfoxide
	DNA	deoxyribonucleic acid
	dNTP	2'-deoxynucleoside-5'-triphosphate (with N = A, C, G, U and T)
	DTM	dithiomethyl-

E

	EDTA	ethylenediaminetetraacetic acid
	equiv.	(molar) equivalents
	ESI(-)	Electrospray-Ionisation in negative mode
	ESI(+)	Electrospray-Ionistaion in positive mode
	EtOH	ethanol
	et al.	et alii

F

	FC	flash chromatography
	FITC	fluoresceinisothiocyanate

	FPLC	fast protein liquid chromatography

G

	G	Guanosine
	[g]	gram

H

	h	hour
	HCl	hydrochloric acid
	HF	hydrofluoric acid
	HPLC	high pressure liquid chromatography

I

	I	iso
	i. e.	id est

K

	KF	potassium fluoride
	KHMDSA	potassium hexamethyldisilylamide
	KOH	potassium hydroxide

M

	m	multiplet
	M	molar
	MALDI	Matrix assisted laser desorption ionization
	mCPBA	3-chloroperoxybenzoic acid
	Me	methyl
	MeOH	methanol
	[mg]	milligram
	min	minutes
	[µl]	microliter
	[µm]	micrometer
	MMT	p-monomethoxytrityl
	mp	melting point
	MTM	methylthiomethyl
	MS	mass spectroscopy

N

	NaH	sodium hydride
	NaOAc	sodium acetate
	NaOH	sodium hydroxide
	NaOMe	sodium methoxide
	NIS	N-iodosuccinimide
	[nm]	nanometer
	NMR	nuclear magnetic resonance
	nt	nucleotide

O

	OD	Optical density
	ODN	Oligodeoxynucleotide

P

	p.a.	*pro analysi*
	PAA	polyacrylamide
	PAGE	polyacrylamide gelelectrophoresis
	PCR	polymerase chain reaction
	PG	protecting group
	pH	potentia hydrogenii
	PP_i	pyrophosphate
	*p*TSA	*para*-toluenesulfonic acid
	py	pyridine

Q

	Q	quartet

R

	R_f	retention factor ("relate to front")
	RP	reversed phase
	rt	room temperature

S

	s	singlet
	sat.	saturated
	SBH	Sequencing by Hybridization
	SBS	Sequencing by Synthesis
	SNP	single nucleotide polymorphism

T

	T	Thymidine
	T	temperature
	t	triplet
	Taq	*Thermus aquaticus*
	TASF	tris(dimethylamino)sulfonium difluorotrimethylsilicate
	TBAF	tetrabutylammoniumfluoride
	TBDMS	*tert.*-butyldimethylsilyl
	tBuOH	*tert.*-butanol
	TDA-1	tris(dioxa-3,6-heptyl)amine
	TEA	triethylamine
	TEAB	triethylammonium bicarbonate
	tert.	tertiary
	TFA	trifluoroacetic acid
	THF	tetrahydrofurane
	*ti*PDSCl	1,1,3,3-tetraisopropyldichlordisiloxane
	TLC	thin layer chromatography
	TMS	trimethylsilyl
	TMSCl	chlorotrimethyl silane
	TOF	time-of-flight
	Tol	toluoyl

U

	U	Uridine
	UV	ultraviolet

8.3 Literature

[1] Watson J. D., Crick F. H., "Molecular structure of nucleic acids; a structure for deoxyribose nucleic acid", *Nature* **1953**, *171* (4356), p. 737-738.
[2] Maxam A. M., Gilbert W., „A new method for sequencing DNA", *Proc. Natl. Acad. Sci. USA* **1977**, *74*, No. 2, p. 560-564.
[3] Sanger F., Nicklen S., Coulson A. R., „DNA sequencing with chain-terminating inhibitors", *Proc. Natl. Acad. Sci. USA* **1977**, *74*, No. 12, p. 5463-5467.
[4] Dunn J. J., Studier F. W., "Complete nucleotide sequence of bacterio-phage T7 DNA and the locations of T7 genetic elements", *J. Mol. Biol.* **1983**, *166*, p. 477-535.
[5] Sanger F., Coulson A. R., "A rapid method for determining sequences in DNA by primed synthesis with DNA polymerase", *J. Mol. Biol.* **1975**, *94*, p. 441-448.
[6] Sanger F., Air G. M., Barrell B. G., Brown N. L., Coulson A. R., Fiddes C. A., Hutchison C. A., Slocombe P. M., Smith M., "Nucleotide sequence of bacteriophage phi X174 DNA", *Nature* **1977**, *265* (5596), p. 687-695.
[7] Smith L. M., Sanders J. Z., Kaiser R. J., Hughes P., Dodd C., Connell C. R., Heiner C., Kent S. B., Hood L. E., "Fluorescence detection in automated DNA sequence analysis", *Nature* **1986**, *321* (6071), p. 674-679.
[8] Takumi T., Fujiwake H., Kurachi Y.," A Dye Terminator Method for Automated DNA Sequencing Using Four Fluorescent Dideoxynucleosides and Thermal Cycling", *Analyt. Sci.* **1997**, *13*, p. 735-739.
[9] Blattner F. R., Plunkett G., Bloch C. A., Perna N. T., Burland V., Riley M., Collado-Vides J., Glasner J. D., Rode C. K., Mayhew G. F., Gregor J., Davis N. W., Kirkpatrick H. A., Goeden M. A., Rose D. J., Mau B., Shao Y., "The complete genome sequence of Escherichia coli K-12", *Science* **1997**, *277* (5331), p. 1432-1434.
[10] Venter, J. C. et al., „The sequence of the human genome", *Science* **2001**, 291 (5507), p. 1304-1351.
[11] Voelkerding K. V., Dames S. A., Durtschi J. D., „Next-Generation Sequencing: From Basic Research to Diagnostics", *Clinic. Chem.* **2009**, *55* (4), p. 641-658.
[12] Swerdlow H., Wu S. L., Harke H., Dovichi N. J., "Capillary gel electro-phoresis for DNA sequencing. Laser-induced fluorescence with the sheath flow cuvette.", *J. Chromatogr.* **1990**, *516*, p. 61-67.
[13] Hunkapiller T., Kaiser R. J., Koop B. F., Hood L., „Large-scale and automated DNA sequence determination", *Science* **1991**, *254*, p. 59-67.
[14] a) Nyren P., Pettersson B., Uhlen M., "Solid phase DNA minisequencing by an enzymatic luminometric inorganic pyrophosphate detection assay", *Anal. Biochem.* **1993**, *208*, p. 171-175.; b) Ronaghi M., Karamohamed S., Pettersson B., Uhlen M., Nyren P., "Real-time DNA sequencing using detection of pyrophosphate release", *Anal. Biochem.* **1996**, *242*, p. 84-89.; c) Ronaghi M., Uhlen M., Nyren P.,"A sequencing method based on real-time pyrophosphate", *Science* **1998**, *281*, p. 363-365.; d) Ronaghi M., "Pyrosequencing Sheds Light on DNA Sequencing", *Genome Res.* **2001**, *11*, p. 3-11.; e) Russom A., Tooke N., Andersson H., Stemme G., "Pyrosequencing in a Microfluidic Flow-Through Device", *Anal. Chem.* **2005**, *77* (23), p. 7505-7511.
[15] a) Roskey M. T., Juhasz P., Smirnov I. P., Takach E. J., Martin S. A., Haff L. A.,"DNA sequencing by delayed extraction-matrix-assisted laser desorption/ionization time of flight mass spectrometry", *Proc. Natl. Acad. Sci. USA* **1996**, *93*, p. 4724-4729.; b) Fu D. J., Tang K., Braun A., Reuter D., Darnhofer-

Demar B., Little D. P., O'Donnell M. J., Cantor C. R., Köster H., „Sequencing exons 5 to 8 of the *p53* gene by MALDI-TOF mass spectrometry", *Nat. Biotechnol.* **1998**, *16*, p. 381-384.; c) Edwards J. R., Itagaki Y., Ju J., "DNA sequencing using biotinylated dideoxynucleotides and mass spectrometry", *Nucleic Acids Res.* **2001**, *29*, e104.

[16] a) Strezoska Z., Paunesku T., Radoslavljevic D., Labat I., Drmanac R., Crkvenjakov R., "DNA sequencing by hybridization: 100 bases read by a non-gel-based method", *Proc. Natl. Acad. Sci. USA* **1991**, *88*, p.10089-10093.; b) Broude N. E., Sano T., Smith C. L., Cantor C. R., "Enhanced DNA sequencing by hybridization", *Proc. Natl. Acad. Sci. USA* **1994**, *91*, p.3072-3076.; c) Drmanac S., Kita D., Labat I., Hauser B., Schmidt C., Burczak J. D., Drmanac R., „Accurate sequencing by hybridization for DNA diagnostics and individual genomics", *Nat. Biotechnol.* **1998**, *16*, p. 54-58.

[17] a) Shendure J., Porreca G. J., Reppas N. B., Lin X., McCutcheon J. P., Rosenbaum A. M., Wang M. D., Zhang K., Mitra R. D., Church G. M., "Accurate Multiplex Polony Sequencing of an Evolved Bacterial Genome", *Science* **2005**, *309* (5741), p. 1728-1732.; b) Mir K. U., Qi H., Salata O., Scozzafava G., "Sequencing by Cyclic Ligation and Cleavage (CycLiC) directly on a microarray captured template", *Nucleic Acids Res.* **2008**, p. 1-8.

[18] a) Braslavsky I., Hebert B., Kartalov E., Quake S. R., "Sequence information can be obtained from single DNA molecules", *Proc. Natl. Acad. Sci. USA* **2003**, *100*, p. 3960-3964; b) Harris T. D. et al., „Single-Molecule DNA Sequencing of a Viral Genome", *Science* **2008**, *106*, p. 106-109.

[19] a) Seo T. S., Bai X., Kim D. H., Meng Q. Shi S., Ruparel H., Li Z., Turro N. J., Ju J., "Four-color DNA sequencing by synthesis on a chip using photocleavable fluorescent nucleotides", *Proc. Natl. Acad. Sci.* **2005**, *102*, no. 17, p. 5926-5931.; b) Ju J., Kim D. H., Bi L., Meng Q., Bai X., Li Z., Li X., Marma M. S., Shi S., Wu J., Edwards J. R., Romu A., Turro N. J., "Four-color DNA sequencing by synthesis using cleavable fluorescent nucleotide reversible terminators", *Proc. Natl. Acad. Sci.* **2006**, *103*, no. 52, p. 19635-19640.; c) Aksyonov S. A., Bittner M., Bloom L. B., Reha-Krantz L. J., Gould I. R., Hayes M. A., Kiernan U. A., Niederkofler E. E., Pizziconi V., Rivera R. S., Williams D. J. B., Williams P., "Multiplexed DNA sequencing-by-synthesis", *Anal. Biochem.* **2006**, *348*, p. 127-138.; d) Turcatti G., Romieu A., Fedurco M., Tairi A.-P., "A new class of cleavable fluorescent nucleotides: synthesis and optimization as reversible terminators for DNA sequencing by synthesis", *Nucleic Acids Res.* **2008**, *36*, No. 4, e25.

[20] Drmanac R., Labat I., Brukner I., Crkvenjakov R., "Sequencing of megabase plus DNA by hybridization: Theory of the method", *Genomics* **1989**, *4*, p. 114-128.

[21] Ohno S., Taniguchi T., "Structure of a chromosomal gene for human interferon β", *Proc. Natl. Acad. Sci. USA* **1981**, *78*, p. 5305-5309.

[22] Khrapko K. R., Lysov Y. P., Khorlin A. A., Ivanov I. B., Yershov G. M., Vasilenko S. K., Florentiev V. L., Mirzabekov A. D.,"A method for DNA sequencing by hybridization with oligonucleotide matrix", *J. DNA Sequencing Mapping* **1991**, *1*, p. 375-388.

[23] Landegren U., Kaiser R., Sanders J., Hood L.,"A ligase-mediated gene detection technique", *Science* **1988**, *241*, p. 1077-1080.

[24] Shamir R., Tsur D., „Large Scale Sequencing By Hybridization", *Proc. of the 5th annual internat. conference on Computational biology* **2001**, p. 269-277.

[25] DeFrancesco L., "Profile: Rade Drmanac", *Nat. Biotechnol.* **2008**, *26*, no. 10, p. 1100.

[26] Heath S. A., Preparata F. P., Young J., "Sequencing by hybridization using direct and reverse cooperating spectra", *Proc. of the 6th annual internat. conference on*

Computational biology **2002**, p. 186-193.
[27] Hyman E. D., "A new method of sequencing DNA", *Anal. Biochem.* **1988**, *174*, p. 423-436.
[28] Nyren P., "Enzymatic method for continuous monitoring of DNA polymerase activity", *Anal. Biochem.* **1987**, *167*, p. 235-238.
[29] Margulies M. et al., "Genome sequencing in microfabricated high-density picolitre reactors", *Nature* **2005**, *437*, p. 376-380.
[30] Shendure J., Ji H., "Next-generation DNA sequencing", *Nat. Biotechnol.* **2008**, 26, No. 10, p. 1135-1145.
[31] Green R. E., Krause J., Ptak S. E., Briggs A. W., Ronan M. T., Simons J. F., Du L., Egholm M., Rothberg J. M., Paunovic, Pääbo S., "Analysis of one million base pairs of Neanderthal DNA", *Nature* **2006**, *444*, p. 330-336.
[32] Metzker M., Raghavachari R., Richards S., Jacutin S. E., Civitello A., Burgess K., Gibbs R. A., "Termination of DNA synthesis by novel 3'-modified-deoxyribonucleoside 5'-triphosphates, *Nucleic Acids Res.* **1994**, *22(20)*, p. 4259-4267.
[33] Wu W., Stupi B. P., Litosh V. A., Mansouri D., Farley D., Morris S., Metzker S., Metzker M. L., "Termination of DNA synthesis by N^6-alkylated, not 3'-O-alkylated, photocleavable 2'-deoxyadenosine triphosphates", *Nucleic Acids Res.* **2007**, p. 1-11.
[34] a) Hovinen J., Azhayeva E., Azhayev A., Guzaev A., Lönnberg H., "Synthesis of 3'-O-(ω-Aminoalkoxymethyl)thymidine 5'-Triphosphates, Terminators of DNA Synthesis that enable 3'-Labelling", *J. Chem. Soc. Perkin Trans.* **1994**, 1, p. 211-217; b) Canard B., Sarfati R. S., "DNA polymerase fluorescent substrates with reversible 3'-tags", *Gene* **1994**, *148*, p. 1-6.
[35] a) Li Z., Bai X., Ruparel H., Kim S., Turro N. J., Ju J., "A photocleavable fluorescent nucleotide for DNA sequencing and analysis", *Proc. Natl. Acad. Sci.* **2003**, *100*, no. 2, p. 414-419; b)Seo T. S., Bai X., Ruparel H., Li Z., Turro N. J., Ju J.,"Photocleavable fluorescent nucleotides for DNA sequencing on a chip constructed by site-specific coupling chemistry", *Proc. Natl. Acad. Sci.* **2004**, *101*, no. 15, p. 5488-5493.
[36] Fedurco M., Romieu A., Williams S., Lawrence I., Turcatti G., "BTA, a novel reagent for DNA attachment on glass and efficient generation of solid-phase amplified DNA colonies", *Nucleic Acids Res.* **2006**, *34*, e22.
[37] Parce J. W., Nikiforov T. T., Burd Mehta T., Kopf-Sill A., Chow A. W., Knapp M. W., "Sequencing by Incorporation", *WO 00/50642* **2000**.
[38] a) Milton J., Wu X., Smith M., Brennan J., Barnes C., Liu X., Rüdiger S., "Modified nucleotides", *WO 2004/018497 A2* **2004**; b) Kawashima E. H., Farinelli L., Mayer P., "Method of nucleic acid sequencing", *EP 1 498 494 A2* **2005**.
[39] a) Seo T. S., Bai X., Kim D.H., Meng Q., Shi S., Ruparel H., Li Z., Turro N. J., Ju J., "Four-color DNA sequencing by synthesis on a chip using photocleavable fluorescent nucleotides", *Proc. Natl. Acad. Sci.* **2005**, *102*, no. 17, p. 5926-5931; b) Ruparel H., Bi L., Li Z., Bai X., Kim D. H., Turro N. J., Ju J., "Design and synthesis of a 3'-O-allyl photocleavable fluorescent nucleotide as a reversible terminator for DNA sequencing by synthesis", *Proc. Natl. Acad. Sci.* **2005**, *102*, no. 17, p. 5932-5937; c) Meng Q., Kim D. H., Bai X., Bi L., Turro N. J., Ju J., "Design and Synthesis of a Photocleavable Fluorescent Nucleotide 3'-O-Allyl-dGTP-PC-Bodipy-FL-510 as a Reversible Terminator for DNA Sequencing by Synthesis", *J. Org. Chem.* **2006**, *71*, p. 3248-3252; d) Bi L., Kim D. H., Ju J., "Design and Synthesis of a Chemically Cleavable Fluorescent Nucleotide, 3'-O-Allyl-dGTP-allyl-Bodipy-FL-510, as a Reversible Terminator for DNA Sequencing by Synthesis", *J. Am. Chem. Soc.* **2006**,

128, p. 2542-2543.
[40] Ollis D. L., Brick P., Hamlin R., Xuong N. G., Steitz T. A., "Structure of large fragment of Escherichia coli DNA polymerase I complexed with dTMP", Nature **1985**, 313, p. 762-766.
[41] Klenow H., Henningsen I., "Selective Elimination of the Exonuclease Activity of the Deoxyribonucleic Acid Polymerase from Escherichia coli B by Limited Proteolysis", Proc. Natl. Acad. Sci. USA **1970**, 65, p. 168-175.
[42] a) Moran S., Ren R. X.-F., Kool E. T., "A thymidine triphosphate shape analog lacking Watson-Crick pairing ability is replicated with high sequence selectivity", Proc. Natl. Acad. Sci. USA **1997**, 94, p. 10506-10511; b) Liu D., Moran S., Kool E. T., „Bi-stranded, multisite replication of a base pair between difluorotoluene and adenine: confirmation by 'inverse' sequencing", Chem. Biol. **1997**, 4, p 919-926.
[43] Pelletier H., Sawaya M. R., Kumar A., Wilson S. H., Kraut J., „Structures of Ternary Complexes of Rat DNA Polymerase β, a DNA Template-Primer, and ddCTP", Science **1994**, 264, p. 1891-1903.
[44] Burgers P. M. J., Eckstein F., "A study of the mechanism of DNA polymerase I from Escherichia coli with diastereomeric phosphorothioate analogs of deoxyadenosine triphosphate", J. Biol. Chem. **1979**, 254, p. 6889-6893.
[45] a) Freemont P. S., Friedman J. M., Beese L. S., Sanderson M. R., Steitz T. A., "Cocrystal structure of an editing complex of Klenow fragment with DNA", Proc. Natl. Acad. Sci. USA **1988**, 85, p. 8924-8928; b) Beese L. S., Steitz T. A.," Structural basis for the 3'-5' exonuclease activity of Escherichia coli DNA polymerase I: a two metal ion mechanism", EMBO J. **1991**, 10, p. 25-33; c) Derbyshire V., Freemont P. S., Sanderson M. R., Beese L., Friedman J. M., Joyce C. M., Steitz T. A., "Genetic and Crystallographic Studies of the 3',5'-Exonucleolytic Site of DNA Polymerase I", Science **1988**, 240, p. 199-201; d) Derbyshire V., Grindley N. D. F., Joyce C. M.," The 3'-5' exonuclease of DNA polymerase I of Escherichia coli: contribution of each amino acid at the active site to the reaction", EMBO J **1991**, 10, p. 17-24; e) Gupta A. P., Benkovic S. J., "Stereochemical course of the 3'->5'-exonuclease activity of DNA polymerase I", Biochemistry **1984**, 23, p. 5874-5881.
[46] Canard B., Cardona B., Sarfati R. S., "Catalytic editing properties of DNA polymerases", Proc. Natl. Acad. Sci. USA **1995**, 92, p. 10859-10863.
[47] Herrlein M. K., Konrad R. E., Engels J. W., Holietz T., Cech D., „3'-Amino-Modified Nucleotides Useful as Potent Chain Terminators for current DNA sequencing methods", Helv. Chim. Acta **1994**, 77 (2), p. 586-596.
[48] a) Summerer D., Rudinger N. Z., Detmer I., Marx A., "Enhanced Fidelity in Mismatch Extension by DNA Polymerase through Directed Combinatorial Enzyme Design", Angew. Chem. Int. Ed. **2005**, 44, p. 4712-4715; b) Strerath M., Marx A., "Genotypisierung – von genomischer DNA zum Genotyp in einem Schritt", Angew. Chem. **2005**, 117, p. 8052-8060; c) Di Pasquale F., Fischer D., Grohmann D., Restle T., Geyer A., Marx A., „Opposed Steric Constraints in Human DNA Polymerase # and E. coli DNA Polymerase I", J. Am. Chem. Soc. **2008**, 130 (32), p. 10748-10757.
[49] a) Strerath M., Cramer J., Restle T., Marx A., "Implications of Active Site Constraints on Varied DNA Polymerase Selectivity", J. Am. Chem. Soc. **2002**, 124 (38), p. 11230-11231; b) Marx A., MacWilliams M. P., Bickle T. A., Schwitter U., Giese B., „4'-acylated Thymidines: A New Class of DNA Chain Terminators and Photocleavable DNA Building Blocks", J. Am. Chem. Soc. **1997**, 119 (5), p. 1131-1132.
[50] a) Kranaster R., Marx A., „New Strategies for DNA Polymerase Library Screening", Nucleic Acids Symp. Ser. **2008**, 52, p. 477-478; b) Kranaster R., Ketzer P., Marx A.,

"Mutant DNA Polymerase for Improved Detection of Single-Nucleotide Variations in Microarrayed Primer Extension", *ChemBioChem* **2008**, *9*, p. 694-697.

[51] a) Ju J., Li Z., Edwards J. R., Itagaki Y., "Massive parallel method for decoding DNA and RNA", *US 2002/0102586 A1* **2002**; b) Ju J., Li Z., Edwards J. R., Itagaki Y., "Massive parallel method for decoding DNA and RNA", *WO 02/29003 A2* **2002**.

[52] a) Kwiatkowski M., "Novel Chain Terminators, The Use Thereof For Nucleic Acid Sequencing And Synthesis And A Method Of Their Preparation", *EP 0 808 320 B1* **1996**; b) Kwiatkowski M., "Chain Terminators, The Use Thereof For Nucleic Acid Sequencing And Synthesis And A Method Of Their Preparation", *US 6,255,475 B1* **2001**; c) Kwiatkowski M., "Compounds For Protecting Hydroxyls And Methods For Their Use", *US 6,309,836 B1* **2001**; d) Kwiatkowski M., "Compounds For Protecting Hydroxyls And Methods For Their Use", *US 2004/0175726 A1* **2004**.

[53] a) Crinelli R., *et al.*, "Design and characterization of deoxy oligonucleotides containing locked nucleic acids", *Nucleic Acid Res.* **2002**, *30*, no. 11, p. 2435-2443; b) Di Giusto D., King G. C., "Single base extension (SBE) with proofreading polymerases and phosphorothioate primers: improved fidelity in single – substrate assays", *Nucleic Acids Res.* **2003**, *31*, e7; c) Di Giusto D.A.. King G. C., "Strong positional preference in the interaction of LNA oligonucleotides with DNA polymerase activities: implications for genotyping assays", *Nucleic Acids Res.* **2004**, *32*, e32; d) Patel P.H., Loeb L.A., "Getting a grip on how DNA polymerases function", *Nature Struct. Biol.* **2001**, *8*, p.656-659; e) Victorova, L. *et al.*, "New substrates of DNA polymerases", *FEBS Letters* **1999**, *453*, p. 6-10.

[54] Metzker M.L. *et al.*, "Elimination of residual natural nucleotides from 3'-O-modified-dNTP syntheses by enzymatic mop-up", *BioTechniques* **1998**, *25*, p. 814-817.

[55] a) Meyer P.R. Matsuura S. E., So A. G., Scott W. A., "Unblocking of chain – terminated primer by HIV-1 reverse transcriptase throught a nucleotide-dependent mechanism", *Proc. Natl. Acad. Sci. USA* **1998**, *95*, p. 13471-13476; b) Knapp D., Ph D thesis, JWG Universität Frankfurt a. M. **2009**, *manuscript in preparation*.

[56] Davoll J., "Pyrrolo[2,3-*d*]pyrimidines", *J. Chem. Soc.* **1960**, *26*, p. 131-138.

[57] a) Saneyoshi H., Seio K., Sekine M., „A General Method fort he Synthesis of 2'-O-Cyanoethylated Oligoribonucleotides Having Promising Hybridization Affinity for DNA and RNA and Enhanced Nuclease Resistance", *J. Org. Chem.* **2005**, *70*, p. 10453-10460; b) Khorana H.G., Tener G.M., Moffatt J.G., Pol H. E., "A new approach to the synthesis of polynucleotides" *Chem. and Ind. London* **1956**, p. 1523; c) Khorana H.G., Razzell W.E., Gilham P.T., Tener G. M., Pol E. H., "Studies on Polynucleotides. XXIV. The Stepwise Synthesis of Specific Deoxyribopolynucleotides (4). Protected Derivatives of Deoxyribonucleosides and New Syntheses of Deoxyribo-nucleoside-3" Phosphates", *J. Am. Chem. Soc.* **1963**, *85*, p. 3821 – 3827.

[58] a) Markiewicz W. T., „Tetraisopropyldisiloxane-1,3-diyl, a group for simultaneous protection of 3'- and 5'-hydroxy functions of nucleosides", *J. Chem Res. (S)* **1979**, p. 24-25; b) Grøtli M., Douglas M., Beijer B., García R. G., Eritja R., Sproat B., "Protection of the guanine residue during synthesis of 2'-O-alkylguanosine derivatives", *J. Chem. Soc., Perkin Trans. 1*, **1997**, p. 2779-2788.

[59] a) McBride L. J., Kierzek R., Beaucage S. L., Caruthers M. H., "Amidine Protecting Groups for Oligonucleotide Synthesis", *J. Am. Chem Soc.* **1986**, *108*, p. 2040-2048; b) Hagen M. D., Chládek S., "General Synthesis of 2'(3')-O-Aminoacyl Oligoribonucleotides. The Protection of the Guanine Moiety", *J. Org. Chem.* **1989**, *54*, p. 3189-3195; c) Janeba Z., Francom P., Robins M. J., "Efficient Syntheses of 2-Chloro-2'-deoxyadenosine (Cladribine) from 2'-Deoxyguanosine", *J. Org. Chem.*

2003, *68*, p. 989-992.
[60] Harwood E. A., Hopkins P. B., Sigurdsson S. Th., "Chemical Synthesis of Cross-Link Lesions Found in Nitrous Acid Treated DNA: A General Method for the Preparation of N2-Substituted 2'-Deoxyguanosines", *J. Org. Chem.* **2000**, *65*, p. 2959-2964.
[61] Craig G. W., Eberle M., Lamberth C., Vettinger T., „Dimethyl Acetylsuccinate as a Versatile Synthon in Heterocyclic Chemistry – A Facile Synthesis of Heterocyclic Acetic Acid Derivatives", *J. Prakt. Chem.* **2000**, *342*, no. 5, p. 504-507.
[62] a) Barnett C. J., Kobierski M. E., „Process for the synthesis of 4-hydroxy-5-halopyrrolo[2,3-*d*]pyrimidine intermediates", *U.S. patent 5,235,053* **1993**; b) Barnett C. J., Kobierski M. E., "A Convenient Method for Regioselective C-5 Halogenation of 4(3*H*)-Oxo-7*H*-pyrrolo[2,3-*d*]pyrimidines", *J. Heterocyclic Chem.* **1994**, *31*, p. 1181-1183; c) Pudlo J. S., Nassiri M. R., Kern E. R., Wotring L. L., Drach J. C., Townsend L. B., "Synthesis, Antiproliferative, and Antiviral Activity of Certain 4-Substituted and 4,5-Disubstituted 7-[(1,3-Dihydroxy-2-propoxy)methyl]-pyrrolo[2,3-d]pyrimidines", *J. Med. Chem.* **1990**, *33 (7)*, p. 1984-1992.
[63] Peng X., Seela F., "Halogenated 7-deazapurine nucleosides: stereoselective synthesis and conformation of 2'-deoxy-2'-fluoro-β-D-arabinonucleosides", *Org. Biomol. Chem.* **2004**, *2*, p. 2838-2846.
[64] Cottam H. B., Kazimierczuk Z., Geary S., McKernan P. A., Revankar G. R., Robins R. K., "Synthesis and Biological Activity of Certain 6-Substituted and 2,6-Disubstituted 2'-Deoxytubercidins Prepared via the Stereospecific Sodium Salt Glycosylation Procedure", *J. Med. Chem.* **1985**, *28 (11)*, p. 1461-1467.
[65] Peng X., Seela F., "Regioselective Syntheses of 7-Halogenated 7-Deazapurine Nucleosides Related to 2-Amino-7-deaza-2'-deoxyadenosine and 7-Deaza-2'-deoxyisoguanosine", *Synthesis* **2004**, *8*, p. 1203-1210.
[66] Holý A., Votruba I., Masojidkova M., Andrei G., Snoeck R., Naesens L., De Clercq E., Balzarini J., "6-[2-(Phosphonomethoxy)alkoxy]pyrimidines with antiviral activity", *J. Med. Chem.*, **2002**, *45 (9)*, p. 1918-1929.
[67] Rolland V., Kotera M., Lhomme J., "Convenient preparation of 2-deoxy-3,5-di-O-*p*-toluoyl-α-D-*erythro*-pentofuranosyl chloride", *Syn. Comm.*, **1997**, *27 (20)*, p. 3505-3511.
[68] a) Seela F., Thomas H., "Synthesis of Certain 5-Substituted 2'-Deoxytubercidin Derivatives", *Helv. Chim. Acta* **1994**, *77*, p. 897-903; b) Seela F., Steker H., Driller H., Bindig U., "2-Amino-2'-desoxytubercidin und verwandte Pyrrolo[2,3-*d*]pyrimidinyl-2'-desoxyribofuranoside", *Liebigs Ann. Chem.* **1987**, p. 15-19; c) Seela F., Westermann B., Bindig U., "Liquid-Liquid and Solid-Liquid Phase-transfer Glycosylation of Pyrrolo[2,3-*d*]pyrimidines: Stereospecific Synthesis of 2-Deoxy-β-D-ribofuranosides related to 2'-Deoxy-7-carbaguanosine", *J. Chem. Soc. Perkin Trans. I* **1988**, p. 697-702; d) Ramzaeva N., Mittelbach C., Seela F., "7-Halogenated 7-Deaza-2'-deoxyinosines", *Helv. Chim. Acta* **1999**, *82*, p. 12-18; e) Seela F., Hasselmann D., Winkeler H.-D., „Synthese von α-Desazaguanosin und Einfluß der Basenkonzentration auf die Phasentransferglycosilierung", *Liebigs Ann. Chem.* **1982**, p. 499-506; f) Zhang L., Zhang Y., Li X., Zhang L., „Study on the Synthesis and PKA-1 Binding Activities of 5-Alkynyl Tubercidin Analogues", *Bioorg. Med. Chem.* **2002**, *10*, p. 907-912.
[69] Okamoto A., Kanatani K., Saito I., "Pyrene-labeled Base-discriminating Fluorescent DNA Probes for Homogeneous SNP Typing", *J. Am. Chem. Soc.* **2004**, *126*, p. 4820-4827.
[70] Nishino S., Yamamoto H., Nagato Y., Ishido Y., "Partial Protection of Carbohydrate Derivatives. Part 19. Highly Regioselective 5'-*O*-Aroylation of 2'-Deoxyribonucleosides in Terms of Dilution - Drop-by-Drop - Addition Procedure",

Nucl. & Nucl. **1986**, *5 (2)*, p. 159-168.
[71] J. Zemlicka, S.Chládek, A.Holy, J.Smrt, "Oligonucleotidic compounds XIV. Synthesis of some diribonucleoside phosphates using the dimethylaminomethylene derivatives of 2',3'-O-ethoxymethylene ribonucleosides", *Collect. Czech. Chem. Commun.* **1966**, *31*, p. 3198-3211.
[72] K. Sonogashira, Y. Tohda, N. Hagihara, „A convenient synthesis of acetylenes: catalytic substitutions of acetylenic hydrogen with bromoalkenes, iodoarenes and bromopyridines" *Tetrahedron Lett.* **1975**, *16 (50)*, p. 4467-4470.
[73] Seela F., Zulauf M., „Palladium-catalysed cross coupling of 7-iodo-2'-deoxytubercidin with terminal alkynes", *Synthesis* **1996**, p. 726-730.
[74] Chinchilla R., Nájera C., "The Sonogashira Reaction: A Booming Methodology in Synthetic Organic Chemistry", *Chem. Rev.* **2007**, *107 (3)*, p. 874-922.
[75] a) Winkeler H.-D., Seela F., "Synthesis of 2-Amino-7-(2'-deoxy-β-D-*erythro*-pentofuranosyl)-3,7-dihydro-4*H*-pyrrolo[2,3-*d*]pyrimidin-4-one, a New Isostere of 2'-Deoxyguanosine", *J. Org. Chem.* **1983**, *48*, p. 3119-3122; b) Seela F., Muth H.-P., "101. 3'-Substituted and 2',3'-Unsaturated 7-Deazaguanine 2',3'-Dideoxynucleosides: Syntheses and Inhibition of HIV-1 Reverse Transcriptase", *Helv. Chim. Acta* **1991**, *74*, p. 1081-1090.
[76] Balow G., Brugger J., Lesnik E., Acevedo O. L., "Positioning of Functionalities in a Heteroduplex Major Groove: Synthesis of 7-Deaza-2-Amino-2'-deoxyadenosines", *Nucl.&Nucl.* **1997**, *16 (7-9)*, p. 941-944.
[77] Taylor E. C., Kuhnt D., Shih C., Rinzel S. M., Grindey G. B., Barredo J., Jannatipour M., Moran R. G., "A dideazatetrahydrofolate analog lacking a chiral center at C-6: N-[4-[2-(2-amino-3,4-dihydro-4-oxo-7*H*-pyrrolo[2,3-*d*]pyrimidin-5yl)ethyl[benzoyl]-L-glutamic acid is an inhibitor of thymidylate synthase", *J. Med. Chem.* **1992**, *35 (23)*, p. 4450-4454.
[78] a) Seela F., Lüpke U., "Mannich-Reaktion am 2-Amino-3,7-dihydropyrrolo[2,3-*d*]-pyrimidin-4-on, dem Chromophor des Ribonucleosids "Q"", *Chem. Ber.* **1977**, *110*, p. 1462-1469; b) Ramzaeva N., Seela F., "88. 7-Substituted 7-Deaza-2'-deoxyguanosines: Regioselective Halogenation of Pyrrolo[2,3-*d*]pyrimidine Nucleosides", *Helv. Chim. Acta* **1995**, *78*, p. 1083-1090.
[79] Gangjee A., Yu J., Kisliuk R. L., Haile W. H., Sobrero G., McGuire J. J., "Design, Synthesis, and Biological Activities of Classical N-{4-[2-(2-Amino-4-ethylpyrrolo[2,3-*d*]pyrimidin-5-yl)ethyl]benzoyl}-l-glutamic Acid and Its 6-Methyl Derivative as Potential Dual Inhibitors of Thymidylate Synthase and Dihydrofolate Reductase and as Potential Antitumor Agents", *J. Med. Chem.* **2003**, *46 (1)*, p. 591-600.
[80] Gangjee A., Devraj J., Queener S. F., "Synthesis and Dihydrofolate Reductase Inhibitory Activities of 2,4-Diamino-5-deaza and 2,4-Diamino-5,10-dideaza Lipophilic Antifolates", *J. Med. Chem.* **1997**, *40 (4)*, p. 470-478.
[81] Patel M., Ko S. S., McHugh R. J., Markwalder J. A., Srivastava A. S., Cordova B. C., Klabe R. M., Erickson-Viitanen S., Trainor G. L., Seitz S. P., "Synthesis and evaluation of analogs of Efavirenz (SUSTIVA™) as HIV-1 reverse transcriptase inhibitors", *Bioorg. Med. Chem. Lett.* **1999**, *9 (19)*, p. 2805-2810.
[82] Buhr C. A., Wagner R. W., Grant D., Froehler B. C.,„Oligodeoxynucleotides containing C-7 propyne analogs of 7-deaza-2'-deoxyguanosine and 7-deaza-2'-deoxyadenosine", *Nucl. Acids Res.* **1996**, *24 (15)*, p. 2974-2980.
[83] a) Breiner R. G., Rose W. C., Dunn J. A., MacDiarmid J. E., Bardos T. J., "Synthesis of new nucleoside phosphoraziridines as potential site-directed antineoplastic agents", *J. Med. Chem.* **1990**, *33 (9)*, p. 2596-2602; b) Bobek M., Kavai I., Sharma R. A., Grill S., Dutschman G., Cheng Y.-C., „Acetylenic Nucleosides. 4. 1-β-D-Arabinofuranosyl-5-ethynylcytosine. Improved Synthesis and Evaluation of

Biochemical and Antiviral Properties", *J. Med. Chem.* **1987**, *30*, p. 2154-2157.
[84] Chang P. K., Welch A. D., „Iodination of 2'-Deoxycytidine and Related Substances", *J. Med. Chem.* **1963**, *6 (4)*, p. 428-430.
[85] Imazawa M., Eckstein F., " Facile synthesis of 2'-amino-2'-deoxyribofuranosyl-purines", *J. Org. Chem.* **1979**, *44 (12)*, p. 2039-2041.
[86] a) Sekine M., Fujii M., Nagai H., Hata T., „An Improved Method for the Synthesis of N^3-Benzoylthymidine", *Syn. Commun.* **1987**, p. 1119-1121; b) Ti G. S., Gaffney B. L., Jones R. A., "Transient Protection: Efficient One-Flask Syntheses of Protected Deoxynucleosides", *J. Am. Chem. Soc.* **1982**, *104*, p 1316-1319.
[87] Pummerer R., „Über Phenylsulfoxy-essigsäure (II)", *Chem. Ber.* **1910**, *43*, p. 1401-1412.
[88] Pojer P.M., Angyal S.J., "Methylthiomethyl Ethers: Their Use in the Protection and Methylation of Hydroxyl Groups", *Aust. J. Chem.* **1978**, *31*, p. 1031–1040.
[89] Butterworth R.F., Hanessian S., "Selected Methods of Oxidation in Carbohydrate Chemistry", *Synthesis* **1971**, p. 70-88.
[90] Zavgorodny S.G., Pechenov A.E., Shvets V.I., Miroshnikov A.I.., "S,X-Acetals in Nucleoside Chemistry. III. Synthesis of 2'-and 3'-O-Azidomethyl Derivatives of Ribonucleosides", *Nucl. Nucl. Nucl.* **2000**, *19 (10-12)*, p. 1977-1991.
[91] Zavgorodny S.G., Polianski M., Besidsky E., Kriukov V., Sanin A., Pokrovskaya M., Gurskaya G., Lönnberg H., Azhayev A., "1-Alkylthioalkylation of Nucleoside Hydroxyl Functions and Its Synthetic Applications: A New Versatile Method in Nucleoside Chemistry", *Tetrahedron Lett.* **1991**, *32 (51)*, p. 7593-7596.
[92] Yoshikawa M., Kato T., Takenishi T., "A Novel Method for Phosphorylation of Nucleosides to 5'-Nucleotides", *Tetrahedron Lett.* **1967**, *No. 50*, p. 5065–5068.
[93] Ludwig J., Eckstein F., "Rapid and Efficient Synthesis of Nucleoside 5'-O-(1-Thiotriphosphates), 5'-Triphosphates and 2',3'-Cyclophosphorothioates Using 2-Chloro-4H-1,3,2-benzodioxaphosphorin-4-one", *J. Org. Chem.* **1989**, *54*, p. 631-635.
[94] Prusiewicz C. M., Sangaiah R., Tomer K. B., Gold A., „A Robust Synthetic Route to 2'-Deoxy-3'-nucleotide Derivatives of Modified Nucleobases", *J. Org. Chem.* **1999**, *64*, p. 7628-7632.
[95] Sun Q., Edathil J. P., Wu R., Smidansky E. D., Cameron C. E., Peterson B. R., "One-pot Synthesis of Nucleoside 5'-Triphosphates from Nucleoside 5'-*H*-Phosphonates", *Org. Lett.* **2008**, *10 (9)*, p. 1703-1706.
[96] Wu W., Freel Meyers C. L., Borch R. F., "A Novel Method for the Preparation of Nucleoside Triphosphates from Activated Nucleoside Phosphor-amidates", *Org. Lett.* **2004**, *6 (13)*, p. 2257-2260.
[97] Burgess K., Cook D., "Syntheses of Nucleoside Triphosphates", *Chem. Rev.* **2000**, *100*, p. 2047-2059.
[98] El-Tayeb A., Qi A., Müller C.E., "Synthesis and Structure – Activity Relationships of Uracil Nucleotide Derivatives and Analogues as Agonists at Human $P2Y_2$, $P2Y_4$, and $P2Y_6$ Receptors", *J. Med. Chem.* **2006**, *49*, p. 7076-7087.
[99] Shiba Y., Masuda H., Watanabe N., Ego T., Takagaki K., Ishiyama K., Ohgi T., Yano J., "Chemical synthesis of a very long oligoribonucleotide with 2-cyanoethoxymethyl (CEM) as the 2'-O-protecting group: structural identification and biological activity of a synthetic 110mer precursor-microRNA candidate", *Nucl. Acids Res.* **2007**, p. 1-10.
[100] Cielak J., Grajkowski A., Kauffman J. S., Duff R. J., Beaucage S. L., „The 4-(*N*-Dichloroacetyl-*N*-methylamino)benzyloxymethyl Group for 2'-Hydroxyl Protection of Ribonucleosides in the Solid-Phase Synthesis of Oligoribonucleotides", *J. Org. Chem.* **2008**, *73 (7)*, p. 2774-2783.

[101] Umemoto T., Wada T., "Oligoribonucleotide synthesis by the use of 1-(2-cyanoethoxy)ethyl (CEE) as a 2'-hydroxy protecting group", *Tetrahedron Lett.* **2004**, *45*, p. 9529-9531.
[102] Földesi A., Keller A., Stura A., Zigmantas S., Kwiatkowski M., Knapp D., Engels J. W.,"The fluoride cleavable CEM group as reversible 3'-*O*-terminator for DNA sequencing-by-synthesis – Synthesis, Incorporation and Cleavage", *Nucl. Nucl. Nucl.* **2007**, *26*, p. 271-275.
[103] Kremsky J. N., Sinha N. D., „Facile Deprotection of Silyl Nucleosides with Potassium Fluoride/18-crown-6", *Bioorg. Med. Chem. Lett.* **1994**, *4 (18)*, p. 2171-2174.
[104] a) Szarek W. A., Hay G. W., Doboszewski B., "Utility of Tris(dimethylamino)-sulphonium Difluorotrimethylsilicate (TASF) for the Rapid Synthesis of Deoxyfluoro Sugars", *J. Chem. Soc., Chem. Commun.* **1985**, p. 663-664; b) Scheidt K. A., Chen H., Follows B. C., Chemler S. R., Coffey D. S., Roush W. R., "Tris(dimethylamino)sulfonium Difluorotrimethylsilicate, a Mild Reagent for the Removal of Silicon Protecting Groups", *J. Org. Chem.* **1998**, *63*, p. 6436-6437.
[105] Saneyoshi H., Ando K., Seio K., Sekine M., "Chemical synthesis of RNA via 2'-*O*-cyanoethylated intermediates", *Tetrahedron* **2007**, *63*, p. 11195-11203.
[106] Keller A., diploma thesis, JWG Universität Frankfurt a. M. **2006**.
[107] Keller A. C., Serva S., Knapp D. C., Kwiatkowski M., Engels J.W., "Synthesis of 3'-*O*-(2-cyanoethyl)-2'-deoxythymidine-5'-phosphate as a model compound for evaluation of cyanoethyl cleavage", *Coll. Czech. Chem. Commun.* **2009**, *74, No. 4*, p. 515-534.

9 Publications and Presentations

9.1 Publications

Földesi A., Keller A., Stura A., Zigmantas S., Kwiatkowski M., Knapp D., Engels J.W., "The fluoride cleavable CEM group as reversible 3'-O-terminator for DNA sequencing-by-synthesis – Synthesis, Incorporation and Cleavage"; *Nucleosides, Nucleotides and Nucleic Acids* **2007**, *26*, p. 271-275.

Provisional U.S. Patent WO/2008/037568; Title: "Reversible Terminators for efficient Sequencing by Synthesis" Inventors: Knapp, Diana Caterina; Engels, Joachim W.; Keller, Angelika; Li, Yangzhou; Gagilas, Julius; Serva, Saulius; Stura, Alina; Földesi, Andras; Estmer Nilsson, Camilla; Kwiatkowski, Marek.

Keller A., Engels J. W., „Synthesis of 3'-O-cyanoethyl-2'-deoxythymidine-5'-monophosphate and its use for the sequencing-by-synthesis technology", *Collection Symposium Series* **2008**, *10*, p. 376-377.

Knapp D.C., Keller A., D'Onofrio J., Lubys A., Serva S., Kurg A., Remm M., Kwiatkowski M., Engels, J. W., "Synthesis of four colors fluorescently labelled 3'-O-blocked nucleotides with fluoride cleavable blocking group and linker for array based *Sequencing-by-Synthesis* applications", *Nucleic Acids Symposium Series* **2008**, *52*, p. 345-346.

Keller A. C., Serva S., Knapp D. C., Kwiatkowski M., Engels J.W., "Synthesis of 3'-O-(2-cyanoethyl)-2'-deoxythymidine-5'-phosphate as a model compound for evaluation of cyanoethyl cleavage", *Collection of Czechoslovak Chemical Communications* **2009**, *74*, No. 4, p. 515-534 (cover story).

9.2 Posters and presentations

9.2.1 Posters

"Synthesis of 3'-O-modified dNTP's and their use for Sequencing-by-Synthesis", ORCHEM 2008, 16. Lecture Conference of the Liebig-Vereinigung für Organische Chemie, September 1-3, 2008 in Weimar/Germany

"Synthesis of 3'-O-Cyanoethyl-2'-dTMP and its use for the Sequencing-by-Synthesis technology", 14th Symposium on Chemistry of Nucleic Acid Components, June 8 – 13, 2008 in Český Krumlov/Czech Republic

9.2.2 Oral presentations

Title of presentation "Synthesis of 3'-modified dTTPs for DNA-Sequencing-by-Synthesis", held on the 2nd ArraySBS project member conference, 01.03.-03.03.06 in

Vilnius/Lithuania

Title of presentation "Synthesis of modified nucleotides as potential reversible terminators for DNA-sequencing by synthesis", held on the 3rd ArraySBS project member conference, 23.08.-25.08.06 in Uppsala/Sweden

Title of presentation "Synthesis of 7-deaza-iodoguanine, 5-iodocytidine and 3'-O-CE-dTMP", held on the 4th ArraySBS project member conference, 08.03.-09.03.07 in Tartu/Estonia

Title of presentation "Synthesis of 7-deaza-iodo guanine and modification of 5-iodocytidine", held on the 5th ArraySBS project member conference, 31.08.-01.09.07 in Frankfurt/Germany

Die VDM Verlagsservicegesellschaft sucht für wissenschaftliche Verlage abgeschlossene und herausragende

Dissertationen, Habilitationen, Diplomarbeiten, Master Theses, Magisterarbeiten usw.

für die kostenlose Publikation als Fachbuch.

Sie verfügen über eine Arbeit, die hohen inhaltlichen und formalen Ansprüchen genügt, und haben Interesse an einer honorarvergüteten Publikation?

Dann senden Sie bitte erste Informationen über sich und Ihre Arbeit per Email an *info@vdm-vsg.de*.

Sie erhalten kurzfristig unser Feedback!

VDM Verlagsservicegesellschaft mbH
Dudweiler Landstr. 99   Telefon +49 681 3720 174
D - 66123 Saarbrücken   Fax   +49 681 3720 1749
www.vdm-vsg.de

Die VDM Verlagsservicegesellschaft mbH vertritt

Printed by Books on Demand GmbH, Norderstedt / Germany